Gadget Consciousness

Digital Barricades:
Interventions in Digital Culture and Politics

Series editors:
Professor Jodi Dean, Hobart and William Smith Colleges
Dr Joss Hands, Newcastle University
Professor Tim Jordan, University of Sussex

Also available:

Gadget Consciousness

Collective Thought, Will and Action in the Age of Social Media

Joss Hands

PLUTO  PRESS

First published 2019 by Pluto Press
345 Archway Road, London N6 5AA

www.plutobooks.com

British Library Cataloguing in Publication Data
A catalogue record for this book is available from the British Library

ISBN   978 0 7453 3533 9   Hardback
ISBN   978 0 7453 3534 6   Paperback
ISBN   978 1 7868 0257 6   PDF eBook
ISBN   978 1 7868 0259 0   Kindle eBook
ISBN   978 1 7868 0258 3   EPUB eBook

This book is printed on paper suitable for recycling and made from fully
managed and sustained forest sources. Logging, pulping and manufacturing
processes are expected to conform to the environmental standards of the
country of origin.

Typeset by Stanford DTP Services, Northampton, England

Simultaneously printed in the United Kingdom and United States of America

Contents

Series Preface

Crisis and conflict open up opportunities for liberation. In the early twenty-first century, these moments are marked by struggles enacted over and across the boundaries of the virtual, the digital, the actual, and the real. Digital cultures and politics connect people even as they simultaneously place them under surveillance and allow their lives to be mined for advertising. This series aims to intervene in such cultural and political conjunctures. It will feature critical explorations of the new terrains and practices of resistance, producing critical and informed explorations of the possibilities for revolt and liberation.

Emerging research on digital cultures and politics investigates the effects of the widespread digitisation of increasing numbers of cultural objects, the new channels of communication swirling around us and the changing means of producing, remixing and distributing digital objects. This research tends to oscillate between agendas of hope, that make remarkable claims for increased participation, and agendas of fear, that assume expanded repression and commodification. To avoid the opposites of hope and fear, the books in this series aggregate around the idea of the barricade. As sources of enclosure as well as defences for liberated space, barricades are erected where struggles are fierce and the stakes are high. They are necessarily partisan divides, different politicisations and deployments of a common surface. In this sense, new media objects, their networked circuits and settings, as well as their material, informational, and biological carriers all act as digital barricades.

Jodi Dean, Joss Hands and Tim Jordan

Acknowledgements

This book was started in one institution and finished in another, so thanks to all my colleagues at my previous institution, Anglia Ruskin University, in particular to Tanya Horeck, Tina Kendall, Sean Campbell and to fellow previous ARU colleagues who are now also elsewhere, Neal Curtis, Seb Franklin, Jussi Parikka, Milla Tiainan and Eugenia Siapera – all of whom have inspired and otherwise helped out over the period when this book has been in development. Many thanks to my colleagues in Media, Culture, Heritage who have all been so friendly and welcoming and make Newcastle University a great place to work. Also specific thanks to Rhiannon Mason, head of the School of Arts and Cultures, Karen Ross, head of research for the school, and James Ash, head of research for Media, Culture Heritage, for supporting such a productive and encouraging environment for research. My appreciation also goes to Pluto Press, David Castle in particular, and to my fellow series editors Jodi Dean and Tim Jordan, with special appreciation to the latter for such a careful reading of the manuscript and a number of insightful comments that have no doubt made this a stronger book. Thanks and love also to my mum, Marilyn, and to her partner Nigel, who have been ever willing to provide respite from the book writing process, and good food and wine! Likewise, thanks and love to my dad, Bryan, and his partner Carola, who have also always been supportive and an inspiration in sticking to doing what they love. Thanks to my good friends Nev and Ibis Kirton, and to Jason Cox, for many years of solidarity. Thanks and love to the McMurtry family: Ruby, Keith and Stuart, who have welcomed me into the fold with such good humour and spirit. The last and the very most thanks and love go to Dawn, who has been with me all the way though the writing of this book – thank you so much!

Introduction

Gadgets are everywhere. Gadgets wake the workforce every morning, organise their days and perpetually punctuate their attention. Evenings are largely filled by more gadgets – streaming movies on 'smart' televisions or surfing YouTube on a tablet, or even face-to-face encounters facilitated by hook-up apps. There is an increasing integration of gadgets with microprocessors, software and ubiquitous networking that places bodies and brains in proximity with these gadgets, but they also open connections with the totality of other gadgets across the planet.

The launch of the iPhone on 9 January 2007 was a watershed moment in the history of gadgets. This device did nothing ostensibly new, there had previously been plenty of devices for directing, communicating, listening to music, entertaining, educating and so forth – but the act of putting them all together in the same device with a touchscreen and intuitive graphical user interface, alongside an ecology of downloadable applications, was new. It made something real that was well known but had only been dimly intuited: the life of a modern human being is a gadget life.

When introducing the new iPhone at the MacWorld 2007 Keynote Address, Steve Jobs told a very excited audience that 'Every once in a while, a revolutionary product comes along that changes everything' (Jobs, n.d.). Under normal circumstances such an announcement could be readily dismissed as marketing hyperbole, but in this instance there was a case to be made. Jobs, who enjoyed a bit of dramatic irony, announced three new products: 'a widescreen iPod with touch controls; a revolutionary mobile phone; and a breakthrough Internet communications device'. Given the widely trailed product launch, Jobs, the audience and everyone concerned was well aware of the joke: 'An iPod, a phone, and an Internet communicator. An iPod, a phone ... are you getting it? These are not three separate devices, this is one device, and we are calling it iPhone.' After this opening gambit Jobs talked the increasingly excited audience through a list of features that, again while not in themselves new – most if not all the elements had actually been developed elsewhere with public funds (Mazzucato 2014) – were integrated and made accessible and 'user friendly' in ways that were new to the consuming public.

Jobs' enthusiasm and well-honed sales pitch made the device seem ever more innovative and ever more desirable – he placed himself half way between the roles of maker and fan. The most immediately striking aspect of the device was the touch screen, which allowed for multiple complex gestures – including point and click scrolling, expanding, shrinking – all of which enabled a fully functional graphical user interface. As Jobs observed, 'we have been very lucky to have brought a few revolutionary user interfaces to the market in our time. First was the mouse. The second was the click wheel. And now, we're gonna bring multi-touch to the market.' Again, multi-touch had been invented elsewhere, by Wayne Westerman doing publicly funded research at the University of Delaware (Mazzucato 2014: 102). Jobs' demonstration indicated just how useful and compelling the multi-touch feature was. Throughout his keynote Jobs performed with the phone, showing the extent to which it was an intuitive and truly multifunctional device. For example, he used a slider icon on the screen to turn the phone off; this apparently minor detail was no accident – it had turned an ordinarily banal action into a sensual connection with the interface. He commented: 'to unlock the phone I just take my finger and slide it across. Want to see that again? ... We wanted something you couldn't do by accident in your pocket. Just slide it across. Boom.'

He continued on to the iPod app in the same manner: 'You can just touch your music, it's so cool!' Again, he enthused about the gestural capacity of the screen and the new scroll function, 'How do I do this? I just take my finger, and I scroll. That's it. Isn't that cool?' Jobs continued demonstrating feature after feature that had previously been scattered across different devices: email, SMS messaging, photography, satellite navigation, note taking, calling, and so on and so on – what had once been technically fiddly and difficult to use and was now rendered simple, intuitive, gestural, integrated. This was a gadget for the twenty-first century; a science-fiction object that had been lurking in the culture for generations had arrived, finally realised.

By the late 2010s these devices have become so standard, so ubiquitous, that they have largely sunk into the background of everyday life. It is something of a revelation to watch Jobs' keynote from 2007 and to witness just how many of the daily gestures and rituals we have enfolded into our bodies were first revealed in that presentation. Over the years the template has been copied across the whole of the sector. Extra features have been added to the devices, better screens, faster connectivity, additional sensors and greater capacity. Other variations of gadgets across multiple companies have proliferated – such as tablets, smart watches and so forth. But despite all that, and even as these devices have spread to become ubiquitous, the core of the idea of a connected multifunctional digital device has remained largely unchanged.

But what really makes such a thing as an iPhone a gadget? This, of course, means asking what exactly we mean by a gadget. It seems culturally and instinctively right to refer to these things as gadgets, and that the contemporary digital devices discussed above count as such. So, for example, we have television programmes such as 'The Gadget Show' in the UK, in which an array of consumer electronics of this kind are showcased and discussed, or a website such as Engadget, that does similar things – albeit at a more serious and technical level. Yet there is more than simply consumer electronics to the gadget. The gadget, as a cultural trope, goes far beyond this and is much older than these sorts of technologies.

The definition given in the Oxford English Dictionary – 'A small mechanical or electronic device or tool, especially an ingenious or novel one' – is helpful but does not seem adequate to quite capture the full sense and use of the term. The image and idea of the gadget has a long and powerful lineage; Leonardo Da Vinci's helicopter designs, for example, present us with a kind of early technical abstract gadget that bestows a new capacity, and does so in very clever way. Variations of the gadget suffuse modern culture – the James Bond gadget is one notable instance: a multipurpose device that comes to the rescue at vital moments. Again, the universe of science fiction bestows many gadgets, including Doctor Who's sonic screwdriver, a device that can be miraculously calibrated to operate on any interstellar technology, again providing the Doctor with the capacity to escape all sorts of tricky situations. The Star Trek hand-held communicator is often cited as an inspirational gadget – as Steve Jobs himself did. There was also Dr McCoy's medical gadget, which could detect and cure just about any medical problem. Best of all was the Star Trek gadget to end all gadgets: the replicator. The common features of all these devices are ingenuity, intelligence, multifunctionality and expanded human capacity.

Such science-fiction gadgets have been matched by any number of real world items; a glance through 'The Sharper Image' catalogue will reveal enough all-in-one cheese toasters, webbed swimming gloves or anti-aging infra-red goggles to keep any gadget lover happy. The more technology advances the closer the science-fiction gadgets get to the real thing. There is clearly an overlap here between genius, novelty, folly, idiocy, uselessness and convenience – but one with world-changing impact when elements of human capacity and connectivity are genuinely enhanced, or at least shifted and recalibrated. In that regard, gadgets have an essential element of progressiveness about them, even if this sometimes entails a hint of folly, or worse. As such, there is always already something slightly sinister, something hidden, in gadgets. We never quite know how they work, and the potential for them to go wrong – to let us down at exactly the worst moment – is ever present. There is always the chance that they harbour a hidden power, ready to brainwash us and control our behaviour – or just reduce us to victims of designed obsolescence, as simple grist for the capitalist mill.

There is also the potential for a doomsday gadget, something beloved of Dr Strangelove and numerous Bond or Doctor Who villains. Such narratives have made their way into much thinking about actually existing gadgets, bringing with them a suspicion that something uncanny and sinister is going on. These concerns are deeply embedded in the cultural psyche and have held sway since the earliest days of techno-logical development. Latent fears manifest as scare stories about mobile phone radiation eating our brains away, or children becoming hopelessly addicted to their games consoles and losing the ability to concentrate or communicate. It is not that these things can't be true, but that they are deemed to be so in the first instance on a cultural and not empirical level.

So why, in a book like this, use the term gadget rather than something more precise? In the first instance I want to talk about the whole techno-logical ecology that surrounds gadgets – this is not a book about iPhones, or the Android operating system or other such technical specifics, but about gadgets as things that we encounter and that shape us as we shape them, not just as individuals but as communities and as a species. If today we usually think of gadgets in the form of a smart digital device of some kind, this is because digital devices have done such a good job of doing what gadgets always did, or aspired to do, that they now set the terms and offer the ideal type. But while the digital device and the gadget have become almost synonymous, they are not entirely coterminous. The richness and ambiguity of the idea of the gadget as progressive is worth holding on to, if for no other reason than to be able to imagine and build different gadgets if the ones we already have become tools in the hands of our enemies or oppressors. Secondly, 'gadget' is a particularly human term and often a term of endearment, referring to things that are loved and used, that are somehow a little bit magical but also deeply practical in multiple ways, and that also fire the imagination – including an element of fear. As such, gadgets are many-sided things that are inevitably a little bit fuzzy around the edges – and it is to this character that I wish to speak.

I am also interested in gadgets because, given their centrality to modern society, economy and culture, what we do with them – as active agents, citizens and workers – will have a profound impact on the future. This means looking at their social and economic place, at how they function in relation to politics and power, and above all in relation to collective thought, will and action.

Another dictionary definition of gadgets includes the idea that they are devices with a purpose, each one being a 'small device or machine with a particular purpose' (Cambridge Dictionaries Online). Purpose is a troubling and contested term in this context. Is purpose in the purview of the person using the gadget, or is it the gadget that directs the purpose of the user? Indeed, is there any such thing as purpose or free will at all? Perhaps there are simply chains of material determination in the hands of history? This is also a central concern of this book. If gadgets are to be an aid to and an enrichment of the human, or indeed precipitate the emancipation of the 'post-human' condition, then the emergence of purpose is a crucial element. We can only properly understand gadgets in relation to that question. Purpose is related to will, and will to intention, which is an expression of consciousness; where will is translated into action for the expression of a purpose it requires self-consciousness. Consciousness is a product of the brain. That is why this book is not only about gadgets but also about how the brain and consciousness operate in and through gadgets to express purpose and take action. Crucially, this is not only, or even primarily, a matter of individual consciousness but concerns our relation to gadgets as social and collective beings, since we, and they, are part of some greater entity or 'assemblage'.

This topic is one that can only be answered, or even further developed, by exploring a further set of core questions about the nature of gadgets, human will, decision making, consciousness, politics and power – and all of these will be explored over the course of the book. In that regard this is not an empirical book about this or that gadget or about the brain and consciousness as such, but a reflection on the ontology of gadgets and their place in human collectivity. With that in mind, and with this world so infused with gadgets, I will be asking what this infusion means for the character of our experience, our development as beings in the world, and indeed for our make up as entities at all, that is, for our onto-logical essence as both distinct 'beings' and part of a broader collective entity. The first three chapters thus address these more philosophical and abstract questions pertaining to gadgets and consciousness, while the latter three work through the implications for politics, activism and the future of society.

Chapter 1, 'The Question Concerning Gadgets', picks up from this introduction to explore the concept of the gadget more fully, both defining and somewhat embellishing the idea – with the aim of providing a more rigorous framework to think with, and including an ethical and normative aspect, but without drifting into a moralising stance. The idea of the gadget as a 'thing' is examined in light of the thought of Martin Heidegger, with more than a small tip of the hat towards his 'question concerning technology'. The important distinction between 'things' and 'objects' is introduced, wherein things are understood to have a more nourishing and profound interconnection with what surrounds them, and the argument is made that technological devices can operate in this way as gadget-things. The chapter further develops this approach by exploring the ways in which gadgets assemble and how we, as users, might interact with them in such a way that we are disposed to them neither as tools of exploitation nor as their willing slaves. This sets up the need to consider the wider context of gadgets as material entities, and as such as part of the political economy of modern societies.

Chapter 2, 'Gadget Materialism', begins with an interrogation of the techno-capitalist framework, or dispositif, within which gadgets are developed and through which we encounter them. The value that gadgets create for capital is explored as an impediment to developing a relationship with them as 'things', and the question around purpose is revisited through the idea of intention and its relation to materiality. This includes a rehearsal of a number of critical perspectives useful for understanding and challenging the manifestation of gadgets as 'objects' within this milieu. The argument is developed that what is needed is a revived focus on intention, in particular in relation to technological change, and that this is politically as well as conceptually necessary. Intention is also a central element and building block in understanding modes of organisation and collectivity in the digital age. As such, a discussion of the nature of action, the political party and the dangers of an over-reliance on technology is undertaken which includes an argument in defence of reflexivity, thought and self-conscious action. This emphasis on intention highlights the need to consider the place of the brain in relation to gadgets, will and action.

Chapter 3, 'Gadget Brain', follows this up by focusing on intention as it relates to the brain and consciousness – asking what the brain can do and how it can interact with gadgets as things. This includes an exploration of the relation between the brain, consciousness and intention, and the issue of the self and self-consciousness, from the perspective of both neuroscience and philosophy. The chapter also addresses the question of collective consciousness and will, and the ways in which brains and consciousnesses interrelate and resonate with each other. It touches on the place of power and capital in relation to the material, with an emphasis on the concepts of plasticity, extended consciousness and class. The question of extended consciousness returns us to the topic of gadgets, setting the scene for the next chapter.

Chapter 4, 'Gadget Consciousness', synthesises and develops the perspectives developed so far, but also ties them to existing formulations of collective and artificial intelligence, drawing a line away from previous approaches to include a reflection on the significance of labour and cooperation in gadget consciousness, including ideas such as multitude and general intellect, and making the point that gadget consciousness is and must be a political as well as a personal and ontological issue. This political angle necessitates a reflection on questions of memory, manipulation and the nature of technological exploitation. This is needed to remain mindful of the sense that gadget consciousness can also be understood as a consciousness flooded and submitted to gadgets as objects (that is, as tools of capital), and as such is subject to the need for the kind of critical theories identified in Chapter 2. This leads to a reflection on the nature of gadget class consciousness, and the possibility that there could be such a thing as gadget false-consciousness. The chapter concludes by pulling the ideas explored together in an examination of the social network Twitter, and asks if the latter can itself be said to 'think'.

Chapter 5, 'Gadget Action', considers the impact of social media in recent upheavals around the world. This issue has been widely debated, but the debates have been primarily on the plane of pragmatic politics, focusing on measurable impacts in 'this' uprising or 'that' protest. While this serves a purpose, there is a serious limitation in this approach, offering as it does only a piecemeal description of correlations of contingent facts, rather than any serious analysis of social change and the underlying collective dynamics. This chapter therefore explores the valences of the move from collective volition to action, and the mode of being, or rather becoming together, that this entails. This includes the development of a distinction between 'resonant' action and 'idiotic' action. The idea of idiotic action addresses the fact that gadgets do not necessarily lead to progressive or emancipatory ends. The analysis draws on Neal Curtis' idea of 'idiotism' as it applies to the neoliberal subject, and explores how digital network culture presents the figure of the idiot not only in the singular but also as the idiot crowd, or mob, akin to Heidegger's notion of the 'they'. Thus, the key question here again is that of the relation of the one and the many, and of subjectivation in relation to gadgets. The chapter will include some illustrative examples of gadget action in which gadgets, supporting social media apps and platforms, have contributed to an evolving of events in different forms.

Chapter 6, 'Gadget Futures', picks up on the themes of idiocy and resonance and explores the openings for distinct gadget futures they entail, from a dystopian future of unfurling control and exploitation to more hopeful variations extrapolating from an orientation towards gadgets as things. One important element is to ask if this requires a commitment to a renewed and revived communism, distinct from any of the actually existing versions that have previously come and gone. The chapter will argue for a view of communism as an opening and a capacity forged out of gadget consciousness, not a prescribed entity. In that sense communism operates here as a marker to designate the need to live and become in common as a condition of the survival of humanity. The chapter develops this question by examining the 'communist hypothesis' and exploring its efficacy for the digital age. It will argue that there are limited chances of 'events' occurring from digital networks alone – where events are understood in the sense of moments that unravel the status quo and reframe the situation – and through which communism may emerge. This means that a 'gadget' communism needs to focus on a dialectical understanding of the digital. In that regard, the prospect of antagonisms internal to networks, and to capitalism itself, need to be worked out and exploited, with the aim of providing a space for a fidelity to a communist ideal across boundaries, while pulling together elements of the digital into a place of compossibility with actions and events outside of the network. Again, the centrality of the links between brains, gadgets and subjectivation will be revisited, and the place of communism in the age of social media posited.

1

The Question Concerning Gadgets

If, as I suggested in the introduction, we can loosely understand gadgets as useful things with a purpose then the question needs to be moved to a more fundamental level. We need to explore the ways in which these things function in our world – how this 'usefulness' manifests itself, how gadgets extend, alter or constitute the framework of the human world. Having already loosely defined gadgets as 'things' we need to start by asking more seriously: what are 'things'? This may seem an obvious question, but it is far from so, and the answer offers some crucial insights into how we can think about and comport ourselves towards gadgets. This requires an address to the nature of things as such and the role they have in orienting human beings to each other and their environment.

There is no way to ask, or have any chance of answering, the question 'what is a thing?' without recognising the foundational contribution of Martin Heidegger. In Heidegger's writing, the 'thing' is initially explored as a concept that is very general: referring to a thing in the world, such as a rock or tree, or more generally to a state of affairs or an activity, or to something which is just 'something (ein Etwas) and not nothing' (1967: 6). But for the purpose of his thinking Heidegger starts from the position of a thing as primarily some-thing in the world, in the first instance 'that which can be touched, reached, or seen, i.e., what is present-at-hand' (1967: 5). In this conception, then, things are outside us, they exist in and of themselves, they are available to our perception and as such are 'present-at-hand' (*vorhanden*). This can also be understood in the alternative translation of Heidegger's original term *vorhanden* as 'objectively present' (2010: 70). So far this offers a view that is fully compatible with answering the questions posed above. Yet according to Heidegger the initial sketching of the thing as objectively present is not sufficient to understand what a thing truly is.

Heidegger argues that science already offers us good workable descriptions of many things – we are able to talk about rocks in terms of their mineral content, of flowers though the study of botany, of computers in terms of electrons and silicon. But there is something about thingness that is not addressed in such positivistic approaches; the question here is not about the specific scientific qualities of this or that thing, because that does not touch on the question of 'what makes a thing a thing and not what makes it a stone or wood; what conditions (*be-dinkt*) the thing' (1967: 8). This is not to discount the scientific description of things that are 'realities, not viewpoints', but to recognise that what we need to focus on is the thing in 'everyday experience' (14).

The focus on everyday experience is in line with Heidegger's broader philosophy in which his understanding of the human is defined in terms of 'being-there' in a specific time and place and in a relation with a particular world, a configuration which is crystallised in the concept of 'Dasein'. Here engagement with the world defines the human, not abstract essences such as 'soul' or 'mind'. As such, to be human is to be active in the world, and so be a part of it – not observing it from a distance or a separate privileged point of view (as subject and object). The human relation to things is to encounter them as part of a world in which we are invested and that we care about. Dasein is always already part of a world that it mutually constitutes.[1]

In his seminal work *Being and Time*, Heidegger frames the thing precisely in this way, drawing on the Greek meaning of 'things' as 'that which one has to do with in one's concernful dealings' (1962: 96). He talks about such things as always being part of a collection, as being what he calls 'equipment', and argues that there is no such thing as 'an' equipment since it always operates as dependent on other equipment: 'To the Being of any equipment there always belongs a totality of equipment, in which it can be this equipment that it is' (97). There is no hammer without a

---

1   I am aware of a number of derivations, critiques and variations of the definition and use of the notion of the 'thing', for example in the work of Bruno Latour, Donna Haraway, Graham Harman, Bill Brown and others. I have chosen to focus on and develop Heidegger's framing here, as this book is not a philosophical intervention in the ongoing debates over the question of the 'thing', but rather a positive argument drawing on a specific concept and its application. The Heideggerian conceptualisation drawn on here enables the discussion of gadgets in a way nuanced enough to recover and retain a notion of conscious human agency, while accounting for the critiques of the philosophy of consciousness. In contrast, the debates derived from the above-mentioned thinkers are broadly framed within a tradition that is moving away from reflection on consciousness towards a post-human or flat ontology rooted in theories of actor networks or affect-driven social relations. This will be discussed further in subsequent chapters, but to address in detail all the possible objections to or developments of the term 'thing' as it is applied here would have required another book in itself.

nail, no nail without a steel works and so forth. Equipment is grasped in the mode of readiness-to-hand.

The notion of equipment thus emphasises the embeddedness of things in concrete world situations, and underlines their centrality to human being-in-the-world. This impacts directly on the way the world is experienced, that is, within a direct practical relationship: only in use is equipment understood as equipment as such. It becomes present to us, revealed, or 'shown to us' only in our 'appropriating this equipment'. We can never grasp it just theoretically, that is 'proximally', but only as being for something: 'equipment for writing, sewing, working, transportation, measurement' (1962: 97). As such, '[t]he kind of Being which equipment possesses – in which it manifests itself in its own right – we call "readiness-to-hand" [*Zuhandenheit*]' (1962: 98), and importantly, '[d]ealings with equipment subordinate themselves to the manifold assignments of the "in-order-to"' (98).

Put simply, we encounter equipment in the practical process of tool use, of doing something with some aim – as with the simple definition of the gadget. It is interesting that the translation of the original German term 'das zeug', can also be rendered as the collective singular 'stuff' as well as the multiple plural 'useful things', as it is in the Joan Stambaugh translation of *Being and Time* (Heidegger 2010: 68). Used as a collective noun, it can also be translated with a word like 'paraphernalia' (see Heidegger 1962: 97, fn 1), highlighting the point that 'useful things' as 'equipment', or even as generic 'stuff', surround us and constitute the environment in which we exist. Useful things permeate our everyday lives and ground our 'world'. For Heidegger, being useful does not simply equate to a tool responding to a discrete instance of human will or volition in a linear temporality of cause and effect, but is something much more profound in the way we engage with the world.

In his later life Heidegger offered a sustained reflection on the nature of the thing. In the cycle of lectures known as *Insight into That Which Is* he argues that things 'gather'. An object is distinct from a thing, and the distinction is as much about the form of relationship, or our comportment towards it, as an inherent quality. We must go beyond simply having something 'before our minds', as with presence-at-hand, because in such a mode 'no road leads to the thinghood of the thing' (2012: 6). If an object is simply touched, reached out for or apprehended then we know it only in a superficial way and we are placed in a relationship with it as subject and object. A purely subject-object relation does a disservice to the 'world' in which the object rests. Heidegger's definition of the thing as thing is highly complex; using the example of a jug, he argues that it is only in its 'presencing' as part of the 'fourfold' that it is revealed as a thing. The fourfold is a concept Heidegger increasingly employs in his later work to refer to the presence of four elements that exist within a thing and that mirror each other within the thing. Depending on the translation, these are: earth, sky (or heaven), divinities (or gods) and mortals. The fourfold division is gathered in things, and all four must be present for a thing to *thing*. Such is the active nature of the thing that Heidegger also uses it here as a verb, so it is that things *thing* when they evoke the fourfold. In the essay 'The Danger', Heidegger tells us that 'World is the fourfold of earth and sky, divinities and mortals' and that 'World lets the thinging of the thing take place' (46).

The four come together, or rather abide in the thing as a doing, and this doing is a gathering. For example, with the jug, it is not in the object itself that its thingness is found but in its pouring, in its use as a process of gathering the elements around it: 'The thing things. Thinging gathers' (2012: 12). Only when the object is integrated into our phenomenological universe, finding its place there, does its thingness become revealed. That is the sense is which what it is to be a thing is to thing (as a verb).

Andrew J. Mitchell, in exploring his translation of the word *verweilen* (to linger or let abide), suggests the idea of abiding is central in that 'It names the way in which the fourfold coalesces in the thinging of the thing' (2012: x). In his essay on 'The Fourfold', Mitchell explores each element in turn, arguing that earth refers primarily to materiality, to 'the material out of which things are composed' (2013: 298). In the case of the jug, this means that it is 'made out of the earth' not in any simple sense (of course) but in a phenomenological mode as 'the way that things exist as sensuously and materially apparent', and this operates with 'shining, phenomenal radiance' because nothing is 'simply inert' (Heidegger 2012: 6).

Sky (or heaven) operates as the space of appearance, and in that sense is a form of medium: '[t]he sky is a space of movement and change' and is vital in the relationality of things, where they 'must enter a space that is capable of receiving them' (Mitchell 2013: 299). In the example of the jug it is the space inside it, ready to receive wine, that constitutes this; however, it is not simply empty space, but space in relation to the material of the jug – held in tension with that. So says Heidegger: 'The empty, this nothing in the jug, is what the jug is as a holding vessel' (2012: 7).

The third of the four is divinity, and '[t]he divinities are messengers' Mitchell tells us. Things are meaningful to us because they have a message, but these messages are here 'hints', '[t]he hint names the presence of what is no longer present' (2013: 299). Thus, there is a play between presence and concealment that the hint reveals – what this means is that we are put in touch which something we can't quite capture, something withdrawn.

Finally, '[t]he mortals are the humans. They are called the mortals because they are able to die' (Heidegger 2012: 17). As Mitchell puts it: '[w]hat is most my own remains outside of me, and this cracks me open, and is thus my fundamental opening to the world' (2013: 300). This has the effect of creating a relationality both towards the other three elements – earth, sky, divinities – but also to other mortals, and so '"mortals" names those beings defined by exposure and openness to the world' (300). This creates a form of togetherness among mortals that produces community. This is evident again in the example of the jug, where part of the essence of the jug as thing is the 'pour', and the practice of thinging includes the gathering of mortals, 'in the gift of the pour that is a libation, the mortals abide in their way' (Heidegger 2012: 11). Thus, Heidegger tells us, '[w]hen we say: mortals, then we already think, in case we are thinking, the other three along with them from the single fold of the four' (17). Things are not merely tools that we use to impose our will; things are lenses though which we view and with which we build our world.

GADGETS ARE THINGS

I began this chapter with the statement that gadgets are things, and in a simple sense this is undeniable; but it is also true in Heidegger's sense of *thing* – or rather, it *can* be true. Modern technological gadgets, however, are certainly not the sort of entity that Heidegger would ever have been likely to have called things, since he had a profoundly negative and sceptical view of modern technology. The possible objection to this categorising of gadgets as things in the Heideggerian sense will be engaged with later in this chapter, but first comes the positive claim.

The first step is to propose that a gadget gathers in a more complex way than a simple thing – it needs more than one function, that is to say it entails a degree of complexity not present in, for example, a hammer, and in that regard it has a capacity beyond the relation of mere tool. This is not to say that a hammer cannot be used for more than one thing; clearly there are many purposes to which a hammer can be put, but within its human social context its primary function (or affordance, to use the more technically useful term) is to hit nails. Hammers, broadly speaking, are discrete pieces of equipment – they play a role, but a singular one. You can add a hammer to a tool kit, but its inclusion in the kit does not really alter the hammer, or the kit; it simply brings them into proximity. A gadget, however, assembles distinct elements that go on to become more than the individual parts. Gadgets are more complex things, in the sense of being both things and equipment at once. They are entities whose multiplicity opens up greater numbers of affordances and connects them to a much greater technical system.

Gadgets have become digital and networked, but at the same time they need to be defined by their bordered character, their thingness, otherwise they simply melt into the broader technical system. That is to say that as well as being integrated into the technical and social web – though wireless connectivity, mobile data systems and so forth – gadgets are things that we carry in our pockets, they are in our cars and backpacks, they are strapped to our wrists and plugged into our eyes and ears. Gadgets have shape, form and material presence; they can be moved about, dropped, broken and displayed.

In defining the gadget as a discrete entity with a border, an inside and an outside, something that can be held and apprehended, we can draw on the idea of a 'technical object'. The philosopher of technology Gilbert Simondon offers a compelling and influential exploration of technical objects. He argues that they evolved from an initial pre-industrial form, akin to the simple tool, towards the mechanised objects of the industrial revolution. Technical objects undertake a process of concretisation in which they become more and more honed, efficient and effective. The evolution of a technical object moves from a number of parts that function together but that can be slotted in or out at different times to improve the working of the device, to something that becomes completely internally attuned, attaining a state of internal resonance in which the whole functions together: 'The technical being evolves through convergence and self adaptation; unifies itself internally according to the principle of inner resonance' (Simondon 2017: 26). For example, in an engine each part develops multiple uses contributing to the whole, the parts increasingly fuse together, becoming overdetermined to the point that each part could not be anything else and nothing else could substitute for that part. Fusing is a movement from the abstract to the concrete; 'one could say', Simondon tells us, 'that the contemporary engine is a concrete engine, whereas the old engine is an abstract engine' (27). Thus we find that the 'technical object exists as a specific type obtained at the end of the convergence series. The series goes from the abstract to the contrary mode: it tends towards a state which would turn the technical being into a system that is entirely coherent within itself and entirely unified' (29).

Simondon's argument certainly offers a helpful way to understand the gadget as a technical object. In the first instance it helps us place gadgets in the category of a discrete entity that can be encountered in the way Heidegger talks of encountering the jug. Secondly, his description captures a very real aspect of the way we encounter gadgets in everyday life; that is, they are most often sold as individual commodities with specific uses and associated patterns of consumption.

While the jug is encountered as a thing in its form of gathering in the 'pour', gadgets can gather according to the affordances and purposes to which they are employed. However, for this to happen, gadgets have to involve a greater level of complication because we encounter them not primarily through immediate physical capacities but via multiple interfaces. In a way we can talk about the jug as an interface between the user and the water, and to an extent the jug mediates between the two – in its bringing together of the fourfold we can interpret its gathering as a mode of mediation. Gadgets however have interfaces in the form of touchscreens, microphones, speakers, clocks, Bluetooth and Wi-Fi, accelerometers and so forth – varying with the specific gadget and its constituent parts. As such, gadgets function as nexus points in multiple interface modes, connecting their users with networks and interlocutors and with other devices, whether through the GPS system or cellular masts, or through other ad hoc local networks via apps such as FireChat. Needless to say, this makes the process of gathering, of the engagement with the fourfold in the thing, a great deal more complicated, but not inconceivable. This means that the mode of gathering of the gadget varies according to the application it is running and the disposition we have towards it at any one time.

For example, early iterations of the iPhone included a spirit level app, which had a pictorial representation of the glass vial one finds embedded in the analogue tool. In practical terms the digital version functioned in the same way as the analogue tool: you would hold the straight edge of the phone against the surface being tested and when the image of the air bubble settled between the two relevant lines on the image of the vial that meant the surface was horizontal. The fact that the iPhone employed a three-axis gyroscope, a sophisticated miniaturised microprocessor, a touch-sensitive screen, complex software and numerous other technical elements did not impact on the experience of the ready-to-handness of the digital spirit level. Just as if the analogue spirit level were to be dropped and smashed, its sudden presencing as nothing but cracked glass and a leaking pool of coloured alcohol would be the same as a broken iPhone emerging from its withdrawn state into an inert block of metal and glass. Similarly, it is equally possible to conceive of these two tools as things. We can picture the craftsman carefully working to make a fine desk, picking the best wood for the job, working with the grain to draw out the most beautiful finish for the desk top – revealing in the way of poesis, as Heidegger would describe it. The spirit level would be an integral part of this process, helping to gather the elements into alignment and ensuring that the desk top remained at precisely the right angle and intersected perfectly with the user as they sit at the desk. One can equally picture the craftsman carefully lining up the desk top with a bubble made of pixels as one made of air. Grasping the thinghood of either seems reasonable in the mode of the thing as it focuses the deep attention of the craftsman in the moment.

The one ineluctable difference is that the gadget, the spirit level app itself, is composed of code and needs to be executed through digital processors, sensors and so on, and rendered ready-to-hand via another device of some sort – in this case the iPhone itself. Of course, the one does not exist without the other; code alone has no meaning on the page, it must be actioned, and can only be such via the material circuitry – thus object and thing exist in parallel in a way that is not the case with the analogue spirit level. It may have an abstract form, that is a design on paper plus several distinct materials from which it is hewn, but this plan is not executable, just as there is also a distinction between the concept of an app, the code that brings it into being and the digital object. As such the digital object (gadget) is always to an extent abstract and concrete simultaneously, in a way that other forms of technical object, such as an analogue spirit level, are not.

This multivalence of the gadget also distinguishes it somewhat from Simondon's definition of a technical object. The specific formation of software and hardware provides for a highly 'individuated' device (i.e., one that is fully contained and 'concretised' in its form), as in the case described above. While the iPhone itself has clearly reached this technical character of individuation, the software layer allows for changes, including unforeseen changes, of use and configuration. The upshot of this is that gadgets, unlike concretised technical objects, can have elements swapped in and out – apps or software instructions, particular sensors or interfaces – which can completely change or reorient a gadget's function. This is the fluid character of the digital and of simulation. So, while the substrate is concretised, the software layer – as a modular system of digital objects – is not. As such, changes to the system's function can be radical according to use, and indeed susceptible to hacks and unrestricted reshaping and reworking.

In this sense, when we are engaging with gadgets as things, we are also engaging with the thingness of the digital. There is a long and well-developed tradition of thinking about the digital in terms of objects, for example in the practice of object-oriented programming (OOP), in which software is developed according to modular objects that are part of programs that can be plucked out, recombined, reused and re-tasked with different capacities and for different purposes. This has the effect of 'black boxing' code into these modules in so far as they are reused and recombined. In that regard such objects or 'instances' of 'classes', as they are also known in OOP, have specific qualities that impact and are significant as essential elements of the kind of digital objects that we encounter directly as users of digital gadgets.

Digital objects are conceivable abstractly in digital terms as objects that can recombine with different devices, machines and systems, with each rendering the object in different but recognisable ways. The digital spirit level could be designed to operate across different devices, to be transmitted across the internet and so on, just as a car moves from a car park to a motorway and onto a ferry. It is only when the digital object comes to be used, is experienced as a tool, and passes through the mode of the ready-to-hand, that it can become a thing, and as such must be experienced as part of a material assemblage. I therefore claim that digital objects are ontologically incapable of being purely digital *things* in and of themselves.

Gadgets then manifest the digital objects in the form of platforms, applications and tools as part of their 'thinging'. There are developmental explorations of such encounters in the early work of Sherry Turkle, who talks about the 'holding power' of computers in relation to the way in which we are captured by digital objects. Discussing video game play, for example, she tells us '[t]he objects in a video game are representations of objects' and that 'a representation of a ball, unlike a real one, never need obey the laws of gravity unless its programmer wants it to' (2005: 69). Thus, such objects are able to capture us in as much as they allow for 'a more perfect expression' of actions. However, Turkle was writing in an era in which the digital, most often accessed through bulky non-portable desktop computers, could still allow for the sense that it was an escape, an alternate universe that meant 'the liberation of the video game from the "real world"' (69). Today, with ubiquitous computing, the real world and its objects have become more fully integrated with handheld devices. The objects on our screens reach through into the materiality of the everyday.

The upshot of this is that we can now think of digital objects as real objects existing in our world, but providing new affordances and forms of holding power. Indeed, one can also say that digital objects are abstractions from real objects, or even entirely new kinds of objects that have parallel predicates to the analogue. We can conceive of digital objects just as any other kind of object, with a set of affordances, roles and functions that are instantiated though the interfaces that mediate between gadgets and users.

However, we should not overlook the fact that gadget interfaces have both manifest and latent elements which embed power. Branden Hookway captures this latent aspect of the interface, explaining that it, 'like the apparatus, is best understood as having emerged out of a dispersed and heterogeneous set of conditions, developments and aspirations', and that it 'bears within itself an accumulation of techniques, technological and scientific developments, and political, social, economic, and cultural inputs and effects'; as such, 'the interface describes a complication and entanglement of power' (2014: 27). This is doubly true at the software layer, where Alex Galloway sees power as shifting from a classical ideological framework – of a concealment of the real conditions of existence under an illusory veil – to a kind of ethic of practice that 'describes general principles of practice' and defines them 'within the context of a specifically human relationship' (2012: 22). In other words, computers, and interfaces in particular, shape the contexts in which we can actually act; it is 'a process or threshold mediating between two states' (22). Again, this is a power issue. The more invisible and 'intuitive a device becomes, the more it risks falling out of media all together' (25). Therefore, we must become mindful of the power embedded in these devices and interfaces.

Being mindful of power does not mean having to become a technical expert in order to exercise a critique. We can still take a mindful attitude without having to break open the black boxes. The affordances, practices of use, effects and affects of gadgets are all open to challenge. In that sense I suggest that there is still a form of power at work here that we can call good old-fashioned ideology. As such, we can reverse engineer the black boxes as enablers and dis-enablers of consciousness, will and action. This kind of reverse engineering is a form of algorithmic account-ability, a variation of ideology critique for the digital age. It is a form of mindfulness that needs to be adopted as part of a general orientation towards gadgets as things.

We now have a definition of the gadget as a device that mediates between the user, the world, other users and other devices – the level of encounter via an interface pertains to a personal or 'meso' level, that crystallises the micro level of constituent elements (the black boxes), such as the material substrate, touchscreen, microchip and digital code. As such we can also claim that the gadget can be conceived as primarily ready-to-hand at that meso level, operating very much as embedded in the experience of everyday life. This applies also to the digital element of modern gadgets, wherein discrete digital objects are part of an assemblage at the meso level that constitutes the gadget as such. In its embeddedness the gadget can be conceived as a thing, if grasped mindfully and contex-tually within a particular practice.

I will then stake a claim that one important aspect of the gadget is its personal scale. This issue of scale is important; in line with the idea of the gadget as *thing* we must recognise that we (as subjects) encounter the gadget right there, as ours, present-at-hand, something which places us in a world and with which we therefore have a very deep and enduring connection. Yet this 'right-thereness' is offset by the gadget's inevitable membership of a broader category, whether that of equipment in Heidegger's sense, or of an ensemble in Simondon's sense, or even what Benjamin Bratton (2015) calls 'The Stack'. These levels all also operate beyond as well as in the gadget – whether conceived as system, equipment, machine or so on, it is always both there and not there, present and in withdrawal. Therefore, fundamental to the gadget as thing is its openness to the outside – to the fourfold – to its mode of thinging (gathering). We can draw across the elements of the fourfold to picture this mode of gathering. From the earth we have the material components that underpin gadgets and connect them together, that can bring a degree of radiance and internal resonance, and that give the thing its own vibrancy.[2] From the sky we have the space of appearance – the electromagnetic waves and networks that allow the appearance and presencing of the medium, that balance the relationality of elements and bring together the 'mortals'. In the divinities we can discern the 'messengers' in the revealing of others, the meanings we draw from our encounters and the 'play of revealing and concealing'. With the mortals comes the recognition of our commonalty in each other's presence and our shared being towards death, with which should come mutual recognition, community, togetherness and solidarity. This conception of the gadget, as bringing the fourfold together in the thinging of the thing, recognises the capacity to produce openness to the world and each other; it allows us to talk about the gadget as thing as a particular entity, giving more substance to its basic definition as a 'small device or machine with a particular purpose', but also seeing it as a thing within a bigger set of relations.

GADGET ENSEMBLES

As well as being discrete personal devices gadgets are parts of more general systems of social relations. This is something more like a machine that goes beyond a personal device and operates on a macro or 'molar' level, which is constituted by combinations of multiple parts. Karl Marx defines machines by their self-perpetuating motive power, which draws on sources of external stored energy, and pushes them to a point where those using them are no longer actually 'using' them, but have themselves become the tools servicing the machines. This is a very useful insight, and is a vision of the machine that starts to entail a 'means of production' within a specific economic configuration. This productive capacity is something that needs to be considered when moving from the personal to the social and political scale. In discussing

---

2 There needs to be an understanding that with the 'earth' there comes a recognition of the fundamental materiality of 'things'. We must also recognise that in gadgets this has its own set of challenges beyond the immediate internal resonance of the gadget in itself – for example, in terms of the use of 'rare earth materials', the extraction and use of which is highly problematic in any number of ways, including labour exploitation. This means that any development of gadgets as things in the long term must address the much broader questions of environmentalism, sustainability, production and so on. The materiality and political economy of gadgets is thus of great importance, and will be discussed more extensively in Chapter 2.

the productive power of capital, Marx was not thinking about the kind of personal devices we have considered so far, but about machines on a grand scale, the result of the scaling up of a centralised factory system. However, Marx's perspective is one that certainly needs contending with when we consider the gadget 'ensemble' as part of the system of capital as it is currently formulated.

The notion of the 'machine' can also offer insights into the gadget when we look to other philosophical quarters. Gilles Deleuze and Félix Guattari were influenced by Simondon's notion of the ensemble, and in their seminal works *Anti-Oedipus* and *A Thousand Plateaus* they develop a concept of the machine that shares much of that character. They tell us that machines are everywhere: 'Everywhere it is machines – real ones, not figurative ones: machines driving other machines, with all the necessary couplings and connections' (1983: 1). These machines are fluid, dynamic and operate at a different scale than industrial machines. Deleuze and Guattari conceive of a machine as a combinatorial entity – a profoundly productive process of concatenation between entities, assembled through linking, coupling and other processes of becoming. Such is the nature of what they call a 'desiring-machine'. They do not see the machinic as mechanical in any traditional sense. As Slavoj Žižek points out, 'What Deleuze calls "desiring machines" concerns something wholly different from the mechanical: the "becoming-machine"' (2012: 14). This is the machine as multiplicity, an assemblage of smaller elements that is unfixed, evolving, becoming; it is not then simply a discrete clearly bordered object, but speaks to some kind of 'system' of elements.

Another useful related concept that puts a more concrete framework around the assemblage is that of the 'dispositif'. This concept has been developed by several recent thinkers – including Michel Foucault, Gilles Deleuze and Giorgio Agamben, amongst others – and though the term has no precise equivalent in English it is often translated as device or apparatus. The usefulness of this concept in relation to the 'macro' level of the gadget is that it includes elements that generally fall outside of a traditional definition of a simple object or tool; it thus allows us to start thinking about broader economic systems, constructions, practices and beliefs – as well as actual devices.

Foucault inaugurates the concept with his definition of a dispositif as 'a thoroughly heterogeneous ensemble consisting of discourses, institutions, architectural forms, regulatory decisions, laws, administrative measures, scientific statements, philosophical, moral and philanthropic propositions' (1980: 194). Deleuze comments that '[i]n the first instance it is a tangle, a multilinear ensemble. It is composed of lines, each having a different nature' (1992: 159). This complements the notion of the machine or 'machinic assemblage'. Foucault also emphasises the function and utility of a dispositif as being a 'formation at a given historical moment ... responding to an urgent need'. This need is one that operates in society as a whole, rather than in terms of individual need, and as such it is defined by a socially dominant 'strategic imperative' (1980: 195). The strategic imperative of the gadget, in the sense in which I am developing it here, is framed by post-Fordism and the proliferation of the social factory; that is, the extension of work into all aspects of life – a claim that I return to in the next chapter.

Giorgio Agamben develops and extends the concept of dispositif:

I shall call an apparatus literally anything that has in some way the capacity to capture, orient, determine, intercept, model, control, or secure the gestures, behaviors, opinions, or discourses of living beings. Not only, therefore, prisons, madhouses, the panopticon, schools, confession, factories, disciplines, judicial measures, and so forth (whose connection with power is in a certain sense evident), but also the pen, writing, literature, philosophy, agriculture, cigarettes, navigation, computers, cellular telephones and – why not – language itself. (2009: 14)

Agamben goes so far as to 'define the extreme phase of capitalist development in which we live as a massive accumulation and proliferation of apparatuses' (15), including the mobile phone. He also announces that he has 'developed an implacable hatred for this apparatus, which has made the relationship between people all the more abstract' (16). While this definition of apparatus is useful in capturing the scope of the molar level of techno-capitalism, of which the gadget is an element, the vast range that it covers risks rendering the term too broad to be meaningful; as such, the addition of the concept of the gadget as a particular element within the ensemble at the meso level is very helpful.

The digital networked gadget is then a thickening, a drawing down of a broader dispositif (as well as an assemblage of the parts from which it is made). It is thus in a sense fractal, in that it captures many aspects of the dispositif of which it is a part, but is also an object in and of itself with its own parts, logic and affordances – a mirror and a miniature of that which it integrates with, but without being only that. At the level of the dispositif we inevitably run into the question of political economy, since it is the context of political economy that constrains how we make and use gadgets and how we are oriented towards them – and that plays a crucial role in the place of gadgets as things that thing (again, I will return to this in the next chapter). However, before making that step, the question concerning gadgets at the meso level requires an acknowledgment of the profound challenge that gadgets also potentially pose.

THE CHALLENGE OF GADGETS

Reflecting on the challenge gadgets pose must take us back to Heidegger. Technology is itself a contested term, but one essential feature of its definition is as a form of relation with the world, that is, as something that affects – or to use Heidegger's term, challenges – the world. When humans start to grasp the world as present-to-hand, as an entity to be understood, broken up, mastered and captured, then there is a danger. The danger emerges when technology is marshalled towards the ends of controlling and dominating its surroundings and users. Dealing with the world in this way, as constituted by *objects* rather than *things*, is what Heidegger sees as the essence of modern technology – or what he calls *Ge-stell* – variously translated as 'enframing' or 'positionality'. In the modern world this mode has come to dominate. This is not, however, simply a question of perception: a 'mere shift of attitude is powerless to bring about the advent of the thing as thing' (Heidegger 2001: 179). This aspect of Heidegger's thought – and not merely of his thought but also of the destructive nature of technology readily visible in the world – raises serious questions about how can we use equipment mindfully without dealing with the world merely as a collection of objects to be stripped of their thinghood.

The Heideggerian scholar Albert Borgmann introduced the notion of 'the device paradigm' to consolidate and explore these ideas (1984: 40). The view of technology that informs the device paradigm is powerful enough that it must be acknowledged and addressed if we are to engage with gadgets constructively. Borgmann develops Heidegger's understanding in such a way that it extends into many technologically developed artefacts (akin to what I am calling gadgets). His programme is here useful in that it shows us one direction in which Heidegger's ideas can be taken, but it is a direction with which I profoundly disagree. Borgmann defines a device as a technology that makes something available to us, such as heat, in a way that is 'instantaneous, ubiquitous, safe and easy' (41). He gives the example of an electric heater that can be activated with the flick of a switch, as opposed to the laborious task of seeking, finding, gathering and preparing firewood, organising it in a fireplace, then tending to the fire to keep it going safely. Such a laborious process is what Borgmann refers to as a 'focal practice', something that captures our attention fully and also draws in others around the focal point – sitting around the fire, telling stories and so forth. Such practices centre on an object that absorbs and gathers, like the fireplace after the making of the fire, or a dinner table after the care of cooking a fresh meal. These actions and objects are about connecting with an entire world and sharing that world with others. Such 'focal things and practices' (196) are associated with Heidegger's conception of the thing, and as such are defined by their practices of 'gathering'.

The device paradigm instigates the very opposite of a focal thing. According to Borgmann, devices dominate in our world of modern technology and create a situation where we are dislocated from our environment and relieved of the skills and knowledge of the quotidian practices that connect us with the world, forgetting that we can actually learn or be given new skills too. Borgmann believes our skills 'are taken over by the machinery of a device' (42). The device is defined only by the particular 'commodity' it provides for us and is abstracted from any further context – in the case of the electric fire, the heat is the commodity provided, without any of the focal activity of building and sharing the fire. Likewise, in the case of cooking, we could consider a microwaved ready meal, in which all that is provided by the device is instant hunger satiation (the food as fodder that fills the stomach is the commodity – in the simplest terms it is that which satisfies a need). Yet such instant food misses out on the attendant pleasures of preparation, care, sharing and enjoyment in the moment. For Borgmann it is the narrowness of this singular purpose of merely generating one commodity that defines a device as a device. Here we have a definition of a device which has some resonance with the idea of a gadget – it is useful – but interpreted in almost entirely negative terms.

Borgmann makes an associated absolute distinction between devices and things. In this division he is clear that the device paradigm applies universally. Any reform or redirection of technology cannot be aimed at any one device or group of devices, because he sees this as only strengthening the device paradigm. Rather, he wants to reorient and limit our relationship with technology as a whole. This is necessary because to admit any capacity to redirect technology towards its use as a focal thing is to lose the critique, so the aim becomes to 'restrict the entire paradigm, both the machinery and the commodities, to the status of a means and let focal things and practices be our ends' (220). This is based on a particular reading of Heidegger that sees the enframing mode of technology as all consuming.

Borgmann therefore recommends that we learn to set our devices aside and use them only to support and enable our focal things and practices. Reform 'requires the recognition and the restraint of the device paradigm, a recognition that is guided by a focal concern' (221). This 'focal concern' thus always takes precedence over the device, leaving the latter as a purely marginal matter: 'Its proper sphere is the background or periphery of focal things and practices' (220).

There is certainly some appeal in this notion of setting aside: it speaks helpfully to the idea of releasement in Heidegger – that we should remove technology from its central role in our lives but still allow it to support our self-development from the margins. Here a good technological life is one lived in parallel with devices, while our actions are directed towards focal things and practices. This includes the aim of automating some areas of life in order to free up time for more focused pursuits; as Borgmann himself puts it: 'one should gratefully accept the disburdenment from daily and time-consuming chores and allow celebration and world citizenship to prosper' (222).

In many ways automation is laudable, but the problem is that in taking this position Borgmann gives up on the possibility of technical reform that would do more than merely improve efficiency or safety and generally support life as it is already lived: a fundamentally conservative position. If devices are forever excluded from the possibility of being focal things then fulfilment or focal activity, and indeed releasement, within or through a device is out of reach. For example, many users of computer games find great satisfaction and fulfilment in them, and there is a tendency towards elitism here which speaks more to the prejudices of the authors concerned than to anything intrinsic about the activities themselves. This attitude can certainly be seen to develop from tendencies in the Heideggerian milieu, but it is an attitude that can and should be set aside because it creates a significant impediment to human growth and evolution, to becoming itself. In this regard, as a specifically 'orthodox' Heideggerian reading of technical devices, Borgmann's position is too rigid.[3]

Even more problematically there is no accounting here for the social or economic inequalities that can make such concerns about focal practices sound rather precious. In concrete terms, a microwave may be a literal lifesaver for a single parent juggling two jobs to survive. To have the choice to set one's devices aside and attend to focal things is a social and cultural luxury that implies an advantaged class position. To make the case for focal things without attending effectively to this kind of exclusion is a significant problem. Even if the argument were to be made that the aim of a just society should be to automate the satisfaction of needs to the greatest degree – so that social wealth, cultural capital and free time were available to all – this would still freeze human development within a specifically essentialist ideological vision of the 'good life'.

A further expression of the nascent class issue within the 'device paradigm' occurs when Borgmann distinguishes wealth from affluence. Borgmann understands wealth as a positive accumulation of opportunity, life skills, comfort, safety and fulfilment – it is homely, 'homely in the sense of being plain and simple but also in allowing us to be at home in our world, intimate with great things' (223). The opportunity for such a life, free from hunger and excessive toil, is 'made possible by technology, and it is centred in a focal concern' (223). Affluence, by contrast, he sees as 'the possession and consumption of the most numerous, refined, and varied commodities' (223). This is presented as a rather sordid thing that is 'confined to technology', and represents an empty acquisitiveness that results in little more than being 'sad and bored'. Again, this distinction is made possible by the belief that we can hold technology at arm's length and orient it towards non-technological goals which then become free of its taint. By definition, any device-centred activity must be acquisitive and aimed at affluence only.

---

3   This is a good place to acknowledge the need to be extremely careful in the adaptation and application of Heidegger's work, given his dreadful political record in supporting the Nazi party in the 1930s and '40s and his failure to truly recognise or apologise for these failings later in life. I cannot here undertake another extensive discussion of this matter, but will note only that my intention, as stated, is not simply to adopt Heidegger's ideas wholesale but to draw on them where he provides valuable and profound insights and to bring them into a context of mutual recognition and respect, employing them where they can be articulated with ideas from different traditions, as I have tried to do here. This is an ongoing tradition going back to, for example, Marcuse's attempt to create a 'Heideggerian Marxism'.

Borgmann tells us that technology 'provides us with the leisure, the space, the books' to achieve great things, because we are spared being 'worn-out by poor and endless work' (223). But what he overlooks is the fact that it is only some people who are spared. Taking his example of books, their contribution to living a good life is incontrovertible – informing, educating, connecting, accumulating knowledge, all in the way of wealth – and Borgmann rightly identifies this as a positive contribution to focal practices. However, it does not require much scrutiny of the publishing industry to reveal practices that do nothing to liberate people from 'poor and endless work': from the indigenous peoples cleared from their forest land in the production of trees for farming, to the heavy toil of workers in paper mills, to the precariously paid and long-hours labour of copy editors and proof readers, to the often unpaid writers and researchers. Not to mention those applications of the book that have been less than 'focal', for example the development of book-keeping in the organisation of mass populations into warring armies, or armies of exploited labour, and so on.

The exploitation problem here is not the result of technology but of the economic system. There is a chain of production and distribution that holds together the object and the commodity (as defined by Borgmann, in this case the reading experience), which means any virtuous 'wealth' can be directly connected to somebody else's affluence and comes at the price of a third party's toil. This is not an attempt to make a distinction between devices and focal things, or technology and tradition, but rather simply to highlight the presence of capital and labour in the background of the political economy of devices. As such, the object appearance of the book as non-technological and autonomous, and as a result its positing as a focal thing, is merely a question of perspective, one which involves a choice to overlook the material systems that support and nourish the world of books.

We can make a similar argument in relation to Borgmann's example of running. He discusses the experience of running as a focal practice that draws the runner into a reflective and connected relationship with the environment that he or she is running through; but while runners thus 'leave technology behind', in the same moment good running shoes 'allow one to move faster, farther, and more fluidly' (221). This distinction is justified by distinguishing between instruments and *devices*: 'technology can produce instruments as well as devices, objects that call forth engagement and allow for a more skilled and intimate contact with the world' (221). Such a distinction affords little justification other than the need to account for the undeniable ubiquity of technology without committing to a totally anti-technological stance.

When extolling some of the virtues he sees in technology Borgmann acknowledges that 'It frees us from the accidental limits of shortness of time, lack of equipment, or weakness of health so that we can turn to the great things of the world in their own right. It frees us for the genuine limits of our endurance, fortitude, and fidelity' (248). Yet this is more than just a side issue – what is being described here is what makes us human. There is always already something inescapably inhuman in the human. Technology affords the 'inhuman' ability to live an 'unnaturally' long life, with the time to learn and dedicate ourselves to particular kinds of mastery. Setting aside or putting down technology in order to appreciate technology is only ever available as an action in light of technology. So while it may be the case that a 'focal practice engenders an intelligent and selective attitude towards technology' (221), it is surely then the place and comportment towards technology, at the heart of everyday life, that is the issue.

I thus claim that focal practices also need to exist within and through gadgets. Gadgets are not things in their essence, but can become so in their design, internal and external relations and orientation. As such we can distinguish gadget-objects from gadget-things. Neither of these orientations is in the essence of gadgets. In this section we have spent quite a lot of time reflecting on the device paradigm, but I believe this is valuable because exploring the critique helps reveal further how we can usefully conceive of the gadget not only as an object outside of us, but also as a part of the human itself.

RELEASEMENT TOWARDS GADGETS

The notion that the human has evolved in tandem with technology is shared by the philosopher of technology Don Ihde, who argues that technology is not only entwined in everyday life but 'supplies the dominant basis for an understanding both of the world and of ourselves' (1983: 10). Ihde places this claim in historical context by arguing that human self-consciousness locks onto the dominant technologies of any period and projects a reflected understanding of the human back onto itself. For example, in the era dominated by mechanical technology our conception of the universe was one shaped by the idea of the clock as an archetypal device, and as such 'the universe itself began to be conceived of as according to a mechanical metaphor' (32) – a view that is also evident in the Cartesian idea of the body as machine. This is part of a structure that is, according to Ihde, 'invariable'. There is a mixture of invariable and variable relations between self and world, and technology provides the ground upon which we construct those relations. While different technological foundations produce different relations to the world, different rituals and so forth, what is invariable is that relations of intentionality and reflection are inflected through technology.

The invariable element is the always-intentional stance humans take towards the world, one that interprets it and builds a sense, both individual and collective, of what that world is and our place in it. It is the content of the self-consciousness that varies according to the specific technologies involved. Thus, Ihde argues that technology actually precedes our scientific understanding of the world; it is not somehow above and beyond the human or human knowledge of the world. This is significant because it means our primary relation to the world is one of praxis: the reflective use of things, hands-on working with equipment, negotiation and intention. In short, our engagement with technology, of which gadgets are clearly a major category, makes our world.

This notion, that technology pre-dates science and the modern world – both ontologically and temporally – indicates that human subjectivity, as defined not by an essence but by a set of relationships with the world and a set of capacities, is malleable. This malleability is also present in the human organ most associated with human subjectivity, the human brain, in relation to which such malleability is commonly referred to as plasticity. Therefore, at its most basic level we have a subjectivity built around a relation between matter – on the one side the human body, brain and sense organs, and on the other technology – and for us this means gadgets.

Since the human is always already entwined with technological devices, the question concerning gadgets hangs precisely on how we interpret the notion of releasement. It does not mean literally a release from technology and devices, but rather the taking of some other kind of comportment towards them, even if this is something still unformed. Of course, this is why it is so hard to identify and to name – specifying the nature of our living in, dwelling in, becoming with or through devices is the challenge. Just what releasement can involve is explored by Heidegger in a number of his later works, and although he makes no definitive and unambiguous statement on the matter it involves at the least some release or rethinking of will. In one of his clearest reflections on the issue, in the 'Conversation on a Country Path About Thinking', Heidegger tells us, 'as far as we can wean ourselves from willing, we contribute to the awakening of releasement'. This is not to encourage a passive relation to the world; it is rather that releasement lies 'beyond the distinction between activity and passivity' and so 'does not belong to the realm of the will' (1966: 61).

This seems a rather ambiguous path, but perhaps it is clearer when thinking about willing as a mode of overcoming and dominating, in alignment with enframing, rather than as being more open in the way of encouraging or developing. The sense of this is also revealed in the discussion of releasement in relation to the disjunction between 'calculative' and 'meditative' thinking. The modern sense of thinking is calculative in that it is directed towards the achievement of specific ends and plans, of representing the world wherein we 'reckon with conditions that are given' (46). This is not necessarily easy thinking – it is necessary and can be complicated – but it is 'not meditative thinking, not thinking which contemplates the meaning that reigns in everything there is'. If calculative thinking is undertaken exclusively, it gives rise to a 'flight-from-thinking' because it dislocates us from our relationship with the fourfold, and is associated with our dealing with objects as present-to-hand. Meditative thinking places us in a different relationship with everything, invoking an openness and questioning as to what surrounds us. As such, '[i]t is enough if we dwell on what lies close and meditate on what is closest' (47).

Part of the idea of supporting the fourfold is inherent in the idea of dwelling. We dwell in as much as we remain connected to the fourfold, as located and grounded, but also as open and ready to reflect and encounter. So it is that we dwell, Mark Wrathall explains, in 'saving the earth, in receiving the sky, in awaiting the divinities, in accompanying mortals' (2005: 113). This is significant in that by dwelling we can step outside of enframing, or positionality, and let things be – in a mode of releasement. So it is that, '[r]ather than forcing everything to be a resource, ... we let it settle into its proper essence. Or if it is already a resource we "secure or shelter it back" to its essence by developing practices that respond to it as something other than a resource' (113). While Wrathall does suggest that this is applicable to non-technological things, we can understand it in the broader sense of technological or calculative thinking, rather than in relation to devices or gadgets as such.

In the end releasement is an ambiguous term, but Heidegger's general aim of challenging the dominance of calculative thinking is clear. It entails that the essence of thinking and action should be neither manipulative nor passive, but rather reflective: 'the essence of thinking can neither be understood as transcendental-horizontal representation nor as calculative thinking; it must be understood as commemorative thinking and attentive reflection' (Denkee 2000: 42).

We can hope to find in technology, and also in gadgets as things, what Heidegger calls the 'saving power'. This requires 'our catching sight of what comes to presence in technology, instead of merely staring at the technological' (Heidegger 1977: 32). This arguably reflects the distinction between the present-to-hand and the ready-to-hand, indicating that a recovery and a recollection of technology as ready-to-hand is, at least partially, where the saving power lies – where technē can still be understood as craft or art. But it also lies in mindfulness and an awareness that the danger of enframing is always there; therein also lies the possibility of truth, and Heidegger reminds us that 'there was a time when the bringing-forth of the true into the beautiful was called technē' (34). The saving power therefore entails questioning – a questioning that opens up alternatives, and that is of course always also a self-questioning. 'The closer we come to the danger', Heidegger tells us, 'the more brightly do the ways into the saving power begin to shine and the more questioning we become' (35).

What would releasement towards gadgets be? It would require accepting their presence and centrality in modern life, but also understanding their double-sided potentiality, represented by the dangers they present in the way of dominating and alienating us, of limiting our connections with nature and with each other when they are delivered to the instrumentality of the object orientation rather than being treated as things. It must mean remembering that we (humans) are not simply rulers of the world, with gadgets as our means of dominating nature and each other. In terms of discerning gadgets as things, this must include a readiness to be open to the revealing that they allow in terms of the fourfold – to the possibility of a gadget being a thing that things (gathers). To do this requires both an understanding and a contestation of the context in which the majority of actually exiting gadgets come into being and are used – namely, capitalism.

2

Gadget Materialism

In the previous chapter I offered an initial framework in which to define and reflect on gadgets. I argued that gadgets can be encountered as either things or objects, and that our orientation towards them should be as things, recognising their potential as focal things that gather. I argued for releasement towards gadgets as a disposition to support this. I also drew a line of connection between gadget, ensemble and dispositif. I touched on aspects of the political and economic formation of gadgets, but in a rather limited way. Therefore in this chapter I will develop these themes with a more contextual and focused exploration of the role of capital, and of power more broadly, in the formation of gadgets, with a particular emphasis on a materialist understanding of both the potential of gadgets but also their limitations as sources of exploitation and alienation. In that regard I will argue that it is capital that constructs the gadget as object, rather than anything in the gadget itself, and as such the question becomes one of finding a balance between determination and conscious-ness in turning away from that path.

THE TECHNO-CAPITALIST DISPOSITIF

Gadgets operate in concrete material conditions. There are significant limitations, controls and sources of alienation and exploitation in their overarching dispositif. As mentioned previously, the dispositif consti-tutes 'a thoroughly heterogeneous ensemble' (Foucault 1980: 194). The dispositif that currently dominates western societies – and much of the wider world in varying degrees of intensity and totality – is that of the techno-capitalism. The techno-capitalist dispositif that we currently inhabit represents 'the extreme phase of capitalist development' (Agamben 2009: 15). A simple definition is offered by Luis Suarez-Villa, who describes it as a form of capitalism 'that is heavily grounded on corporate power and its exploitation of technological creativity' (2009: 3). The term techno-capitalism was coined by Douglas Kellner (1989:

177), who understands our current technological age as a specific config-uration of capitalism, and argues that, despite the contentions of certain postmodern, post-industrial and post-Marxist thinkers, it remains the case that

> capitalist relations of production and the imperative to maximize capital accumulation continue to be central constitutive forces … commodity production and wage labor for capital still exist as funda-mental organizing principles, as does the control of the economic by a corporate elite, the exploitation and alienation of labor, production for profit rather than use, and capitalist market, exchange relations. (177)

This techno-capitalist dispositif is held together by the logic of capital accumulation, and indeed, in times of increasingly challenging conditions for profit-making, it is this imperative that explains the development of technology away from satisfying human needs towards capturing intellect and marginalising labour. Even as far back as 1989 Kellner recognised a 'configuration of capitalist society in which technical and scientific knowledge, automation, computers and advanced technology play a role in the process of production parallel to the role of human labor power, mechanization and machines in earlier eras of capitalism' (178). Gadgets operate in the context of capital by necessity in a way that bends them to its purpose.

From the viewpoint of the techno-capitalist dispositif, gadgets are useful five times over. Firstly, as commodities in and of themselves: they are material entities produced within a factory system, much like cars or refrigerators before them. They are functional, exchangeable in a market place, and most certainly objects in as much as they are produced and circulated as such. Secondly, they are also platforms for the production and circulation of further digital commodities. As such they function as facilitators of another set of exchanges that engulf entire networks of labour, circulation and realisation. All these transactions occur via further digital objects hosted by gadgets, in the way of applications that facilitate online entertainment, games, social networking and so forth. Thirdly, gadgets provide a means for another layer of commodity exchange in the form of facilitating further services independent of the gadget itself: food deliveries, taxis, dating and so on. Their fourth use concerns their utility in terms of control and management. Numerous apps have been created to organise, regiment and micro-manage labour down to the 'fractal' level. We see the total monitoring of workers in organisations such as Uber or Deliveroo, whose apps bind drivers to their cars and riders to their bikes, monitoring their performance and their level of 'customer service' to the smallest degree. Here, specially designed apps work in tandem with the broader imperatives and legal structures of neoliberal capitalism. Finally, there is the value generated on the consumption side of the economy. This further layer of value creation and extraction in the political economy of gadgets lies in user-generated content and data. User-generated content can take the form of general social interaction, mined for data and monetised, or indeed the value extracted from all other forms of social production in the mode of what Tiziana Terranova coined as 'free labour'. Writing as far back as 2000 Terranova told us: 'the Internet is animated by cultural and technical labor through and through, a continuous production of value that is completely immanent to the flows of the network society at large' (2000: 34); in societies with highly active, knowledgeable consumers of culture, '[f]ree labor is the moment where this knowledgeable consumption of culture is translated into productive activities that are pleasurably embraced and at the same time often shamelessly exploited' (37).

The prevalence of such value creation and extraction has become ever clearer in the succeeding years. Social media platforms and the shift to the so-called Web 2.0 have colonised and transformed so much more of the economy than was envisioned even by Terranova and the Italian autonomist thinkers who inspired her (more on whom below).

The issue of control is also very much present in civic life too, with some of the most trenchant critiques of gadgets, or at least of the social media that flows through them, centring on the control and manipulation of consumption and of populations, or so called biopower. The storing of data, the surveillance of personal preferences and communications, the nudging of opinions and behaviour by those in control, have evolved into a form of neuro-programming. The intimate relations we have with our gadgets means that personal perspectives can be quantified and turned into data sets of political views, opinions and beliefs, which can be modulated by false or manipulative information introduced into the public sphere.

Such is the impact of these changes there have been claims that we are moving into a new iteration of capitalism and democracy, beyond techno-capitalism towards a platform capitalism. Nick Srnicek argues that capitalism's response to the financial crisis of 2008 produced a dual effect. Firstly large amounts of capital were created, generated by central bank policies of quantitative easing that needed investment opportunities; secondly, alongside that, computer processing power and data-gathering techniques were vastly increased, which meant '[m]assive new expanses of potential data were opened up, and new industries arose to extract these data' (Srnicek 2017: 40). Srnicek argues that this happened to such an extent 'that data would become the raw material to jumpstart a major shift in capitalism' (41). This shift has become manifest, he claims, in a new kind of firm: the platform (42). Platforms operate according to a specific set of characteristics: they 'intermediate between different user groups'; they take advantage of network effects to develop monopoly tendencies 'by having a designed core architecture that governs the interaction possibilities'; and thus, by 'providing a digital space for others to interact in', they 'position themselves so as to extract data from natural processes' (48).

Whether this is a fundamentally new stage of capitalism is debateable, but we may not need to see it as such for this perspective to still be helpful. We can, rather, see it as a new constellation, in the same way as Douglas Kellner deemed techno-capitalism a new constellation and not a new form. As a new constellation of capital, the platform variation is also a double-sided phenomenon of both 'progress and domination' (Kellner 1989: 182), and should be understood as such.

Part of this new constellation is arguably an extra layer of value that Matteo Pasquinelli describes as '*network-value*' (2009: 7), which is captured according to what he calls '*cognitive rent*' (10). Thought of in this mode, the gadget in many ways brings forward and intensifies the long-understood link between production and consumption. Marx, in the *Grundrisse*, makes this link clear when he tells us: '[p]roduction, then, is also immediately consumption, consumption is also immediately production. Each is immediately its opposite. But at the same time a mediating movement takes place between the two' (1973: 91). In Marx's understanding, while the two moments are intrinsically connected they nevertheless remain distinct; a dialectical tension holds or folds them together, but even if brought into mutual reliance in abstraction they retain this 'mediating movement'. Marx also clearly sees in this dialectic the moment in which the subject of the consumer is created. Production 'produces not only the object, but also the manner of consumption, not only objectively but also subjectively. Production thus creates the consumer' (92). This link between production and consumption is one of the fulcrum points of any critical theory of capitalism, in which the relation of the two moments must be challenged, given that the subject's self-realisation as a producer in the moment of consumption is vital to their capacity to resist reification and commodity fetishism.

However, the gadget creates a new dynamic in which this tension becomes so compressed as to be become almost indiscernible. It operates simultaneously in a triadic state: between the immediate modes of consumption and production it also enfolds the moment of subjectivation. It is this indiscernibility that accounts for some of the utopian discourses of digital capitalism as well as cyber-utopianism – the belief that we have somehow transcended the relation of capital and labour, that exploitation and alienation have been overcome. In all these ways, the gadget is becoming the commodity par excellence that instantiates the techno-capitalist dispositif and makes it an immanent presence in everyday life. This is a consciousness that is framed, contained and modulated by gadgets. It is the gadget as a limitation and yoke of our material conditions. In the sense just described, the current human condition is largely determined by the gadgets within the techno-capitalist dispositif; this is very much the gadget as object. To challenge this situation, we first need to instigate a critique of this condition before moving to a more positive vision.

CRITIQUE OF THE GADGET-OBJECT

There is a significant and varied body of critical theory on which we can draw to strengthen the critique of gadget-objects and support the advocacy of gadget-things. This goes beyond the Heideggerian perspective to draw on the rich Marxian materialist tradition. These two perspectives can be seen as contradictory but, as I will argue, the Marxian perspective has a lot to offer here.

One helpful concept from critical theory is that of reification: the objectification, commodification and instrumentalisation of the human. This is a concept Axel Honneth has recently recovered and reworked, drawing on a number of other thinkers in the tradition of critical theory. Honneth defines reification in terms of human relations that are characterised by 'cold, calculating purposefulness' and that reflect 'an attitude of mere instrumental command', creating a situation in which even the human's inner life is 'infused with the icy breath of calculating compliance' (2008: 17).

As critique, the theory of reification insists on a view of the human wherein there ought to be a rich inner life and at least some degree of free thought, association and action, balanced against the cruelty of objectification and commodification. This belief in free action has its roots in the Marxian notion of species-being – the idea that human beings are characterised by their capacity to shape their own conditions of existence. Species-being is not an essentialist claim about the nature of the human, but precisely the opposite, premised on the understanding that existence precedes essence. Reification is thus a mechanism that deviates from the given condition of the human being as open and in a condition of perpetual becoming. Honneth extrapolates from György Lukács' conception of reification as a general social condition within which we are subsumed: 'reification constitutes a multi-layered and stable syndrome of distorted consciousness' (25), and Lukács 'sees all members of capitalist society as being socialized in the same manner into a reifying system' (26).

While this view is plausible it is necessary to recognise that this negative aspect is not the final word, that the critique still harbours an emancipatory potential. Emancipation from reification is found in the potential for a 'proper human praxis' (26) and the 'normative challenge' it contains. So, this is not about decay from a pure form of moral action but rather concerns the empirical imposition of a condition that needs to be overcome.

Honneth's reading of Lukács gives us a foundation for understanding, critiquing and, crucially, finding a way beyond the conception of the gadget as purely an element of the techno-capitalist dispositif. This is not to deny that it is such, but rather to find an opening whereby it can have another function, not only on a personal level but also more importantly at a social and political level. Honneth sees in Lukács a vision of a non-reified praxis that possesses 'the same characteristics of empathic engagement and interestedness that have been destroyed by the expansion of commodity exchange' (27). For Honneth, Lukács offers a minor thought wherein 'he judges the defect of reifying agency against an ideal of praxis characterised by empathetic and existential engagement' (29).

Honneth develops this point with an appeal to the resonances between Lukács analysis and Heidegger's concept of care. The argument is that, concealed in our everyday relations with the world and ourselves – whether in the reified commodity form that structures our relations, or in the alienated orientation that is the present-to-hand – lies a true form. In Heidegger that true form is 'care', which is never totally expunged in modern life and in which 'pre-reflective knowledge or marginal practices remain present in such a way that critical analysis could make us aware of it at any time' (33). In Lukács this is found in the idea of an engaged praxis in which such a caring stance would be actualised. Such a stance, I would add, entails another point of resonance: that of a mindful orientation towards the 'thing'. This mindful orientation is supported by Honneth's reading of the two thinkers' positions in terms of subject-object relations. He argues that with these concepts we have not just a simple communicative action between interlocutors, but a stance that includes the self and the world as well as other subjects. Honneth unifies these positions by suggesting that his conceptualisation of recognition captures them both: they share 'the notion that the stance of empathetic engagement in the world' arises 'from the experience of the world's significance and value', and that a 'recognitional stance therefore embodies our active and constant assessment of the value that persons or things have in themselves' (38).

Honneth sees recognition as preceding, and taking priority over, cognition. He bases this argument in developmental psychology, citing evidence from studies that indicate the necessity of emotional bonding and identification for subsequent cognitive reciprocity and second-person perspective taking. On this basis, he can then extrapolate to the connections between Lukács and Heidegger over the question of care and its alignment with recognition. Thus the claim is that 'our efforts to acquire knowledge of the world must either fail or lose their meaning if we lose sight of this antecedent act of recognition' (47). This provides a baseline of recognition that can then be understood as that which is lost or distorted by reification. This entails in effect an attempt to reconstitute and revivify the lifeworld. The need is then to think about the revivifying of care through the gadget, which I will return to below.

There is a related though distinct approach that also offers useful insights for a critique of the gadget-object. Félix Guattari offers a profound challenge to what he defines as Integrated World Capitalism in his late work *The Three Ecologies* (2000). He tells us: 'Post-industrial capitalism, which I prefer to describe as Integrated World Capitalism (IWC), tends increasingly to decentre its sites of power, moving away from structures producing goods and services towards structures producing signs, syntax and – in particular, through the control which it exercises over the media, advertising, opinion polls, etc. – subjectivity' (2000: 47). This can clearly be understood as another version of the instrumentalisation of the subject, wherein subjectivity itself becomes increasingly integrated into the economy, and where surplus time also becomes part of this process. Guattari notes that, '[t]hrough the continuous development of machinic labour, multiplied by the information revolution, productive forces can make available an increasing amount of time for potential human activity. But to what end? Unemployment, oppressive marginalization, loneliness, boredom, anxiety and neurosis?' (27)

While this view differs from the more obviously Marxian notion of reification, and thinkers like Guattari are often cited as post-Marxist, there are nevertheless a great number of commonalities. For example, Guattari describes the 'dominant modes of valorizing human activities' of this technologically dominated society as 'those of the imperium [Latin: "authority"] of a global market that destroys specific value systems and puts on the same plane of equivalence: material assets, cultural assets, wildlife areas, etc.' (29). The rendering of difference into equivalence is of a piece with the definition of reification, although Guattari offers a very different understanding of the human. Nevertheless, the diagnosis of technology as a levelling and destructive force is present. Although writing before the age of the smartphone and the app, Guattari offers prescient insights for the analysis of gadgets. We can see this in his view of the 'ecological register' of human subjectivity as an area of control by the forces of capitalist mass media and subjectivation, but also in his suggestion that we look 'into what would be the dispositives of the production of subjectivity, which tends towards an individual and/or collective resingularization, rather than that of mass media manufacture, which is synonymous with distress and despair' (34).

The relevance of Guattari's work can perhaps be seen even more clearly in the contemporary work of the thinkers he has influenced, which includes those members of the tradition of 'workerism' or, as it is also known, 'Autonomist Marxism' – for example, Franco 'Bifo' Berardi and Maurizio Lazzarato, who have both written extensively on the combined effects of informational capitalism and digital technology and drawn on Guattari's work. In several of his works Berardi focuses particularly on the way that human subjectivity comes under stress and is reshaped by new forms of digital labour and their expansion into all aspects of life. In his book *The Soul at Work* (2009) we are told that in the digital economy – in which information processing, creative and affective work are the dominant forms of labour – cognition and subjectivity become the sources of value, and as such are increasingly the target of capture by capital. Indeed, it is the brains of such workers that become the targets of integration into the broader dispositif of capital: 'Info-workers can be seen as neuro-workers. They prepare their nervous system as an active receiving terminal for as much time as possible' (90). As such, the 'function of command is no longer a hierarchical imposition, localized in the factory, but a transversal, deterritorialized function, permeating every fragment of labor time' (88). It is here that the relevance to the critique of gadgets can be detected, in this case in the form of the mobile phone, which Bifo sees as realising 'the dream of capital: that of absorbing every possible atom of time at the exact moment the productive cycle needs it' (90). This vampirism, according to Bifo, inflicts serious damage to the mental well-being of all concerned. He tells us that there is now a 'factory of unhappiness' in which there is 'a sort of permanent electro-cution producing a constant mobilisation of psychic energy' (97), and identifies 'swarms of cognitive workers more and more affected by psychopathological syndromes and stress' (98).

This reliance on gadgets such as mobile phones for work is also reflected in their omniscience in the production of forms of subjectivity – the core component in the affective economy. Deleuze and Guattari develop a concept of 'machinic enslavement' that accounts for the integration of human beings into the machine of production, as opposed to merely being subjects of the machine, which entails 'processes of normalization, modulation, modelling, and information that bear on language, perception, desire, movement, etc., and which proceed by way of microassemblages' (Deleuze and Guattari 1987: 458). Maurizio Lazzarato builds on this idea and explores the development of a capitalist subjectivity within the confines of machinic enslavement. He tells us that '[m]achinic enslavement dismantles the individuated subject, consciousness, and representations, acting on both the pre-individual and supraindividual levels' (2014: 12). This is a view of subjectivity which is distinct from the classic enlightenment understanding of the singular unified subject, from which the process of alienation would estrange us. It produces a far more fragmented and multiplicitous entity: 'machinic enslavement takes in a multiplicity of modes of subjectivation, a multiplicity of states of consciousness, a multiplicity of unconsciousnesses, a multiplicity of realities and modes of existence' (90). The notion of machinic enslavement offers a perspective more in keeping with the manifold character of contemporary subjectivity than the classical Marxian conception of oppression and alienation. Yet even so, Lazzarato still insists on the term 'enslavement'. By this he refers to our capture by machines: 'Humans-machines relation are always on the order of a coupling, an assemblage, an encounter, a connection, a capture.' In Marx's time the kinds of machines humans encountered were primarily in the workplace, while outside the factory subjects were limited to apparatuses such as the railways; yet now machines 'are everywhere and especially in our daily lives' (91). Lazzarato describes the daily rituals and patterns of behaviour, from our morning alarms to our entertainments, communications systems and modes of consumption, that draw us into this enslavement.

This echoes Bifo's thinking, who in *The Uprising* tells us that 'In a hypercomplex environment that cannot be properly understood and governed by the individual mind, people will follow simplified pathways and will use complexity-reducing interfaces' (Berardi 2012: 15). Such an imposition and reaction is very much in line with the way I have described the confrontation with the gadget in object mode, but here its tool-like manifestation is precisely one that divides, orders and redirects the subject into a mode of operation whereby the machinic assemblage, which includes the user, pursues a pathway already inscribed by the orientation of the code, processing the 'dividuals' though which it flows. There is little scope here for a moment of stepping back, of reflection, of gathering – only a headlong plunge into the prescribed 'work-flow' of the situation.

Yet Lazzarato does see a possibility for escape in this kind of subjectivation: not moving back to recapture a past form of subjectivity, but rather embracing the element of 'deterritorialisation', refusing to allow capital to 'reterritorialise' in the way of 'manufacturing an individuated subject adapted to the dominant significations that assign him a role, an identity, and a function within the social division of labor' (2014: 92). This perspective echoes the anti-reification position of Honneth, though it envisions a very different form of subjectivity undertaking the action. While Honneth's subject is still the individual pushing back against the incursion and reclaiming autonomy from capital through means of care, Guattari, Lazzarato and Bifo's subject is engaged in a process of becoming deterritorialised. While in very different registers and traditions, the underlying impulse here is the same – to reclaim something subsumed by capital and to remove it from the value chain, either via care in the case of Honneth, or the openness of deterritorialisation in Guattari. Both still contain traces of Marx, and of aspects of the notion of species-being, in respectively emphasising care and becoming.

While there is some hope offered in these notions of care and becoming, another variation of critical thought – which comes from a psychoanalytically influenced Marxism – gives us pause to wonder if even these are simply another way in which we are captured by gadgets. Jodi Dean (2009, 2010) offers a critique of what she calls 'communicative capitalism', understood as a manifestation of capitalism's latest informational, post-industrial phase. The concept takes communication as the central activity upon which capital now relies, both to generate value and to maintain control. This is not a question of communication in the traditional sense of either propaganda or instruction, though of course both still take place, but rather concerns the fact that the time of users is increasingly drawn into the creation and circulation of messages. In Dean's analysis, the major problem with this is that the messages are always already subsumed by capital, with little chance of any intersubjective understanding being achieved; that is, the messages are never actually received, which effectively nullifies the use of gadgets for collective agency.

Dean illustrates the point with reference to the period prior to the invasion of Iraq by a US-led coalition in 2003, which was widely opposed around the entire world. This generated a vast amount of public discussion, much of it conducted online, and resulted in the huge anti-war demonstrations of 15 February 2003. While this has been understood as an example of the potential of digital communications to bring about mass resistance and protest, including by myself (Hands 2006), Dean sees it as indicating precisely the opposite: 'The terabytes of commentary and information ... did not indicate a debate over the war ... the anti-war messages morphed into so much circulating content, just like all the other cultural effluvia flowing through communicative capitalism's disintegrated spectacles' (2009: 20). The endless circulation and multiplication of the message drains it of meaning and leaves it disconnected from any effect. This is reflected, according to Dean, in the fact that the invasion of Iraq went ahead regardless, and is indicative of a general failure of the public sphere, where discourse occurs in a vacuum and can only offer 'new slogans, images, deflections and attacks' (21). A situation in which contributors talk past each other means a lack of 'common terms, points of reference or demarcated frontiers' (22).

The multiplication of opportunities to communicate via ubiquitous devices produces a situation where 'the deluge of screens and spectacles coincides with extreme corporatisation, financialisation, and privatisation across the globe. Rhetorics of access, participation, and democracy work ideologically to secure the technological infrastructure of neoliberalism' (23). In other words, gadgets are unavoidably contributing to the sustenance of an exploitative, isolating and alienating form of life in which discourse of all kinds shifts into a commodity form, regardless of its content. The production of words as labour in the social factory means that the 'morphing of message into contribution is a constitutive feature of communicative capitalism' (26).

In the framework of psychoanalysis this condition amounts to a fatal decline in 'symbolic efficiency'; that is, in the capacity for message and meaning to transfer across contexts and situations: '[t]here is no longer a Master signifier stabilising meaning, knitting together the chain of signifiers', and the result is 'unbearable suffocating closure' (Dean 2010: 6). Symbolic efficiency online has now declined to the point that messages miss their target, are not understood or acted upon, and yet there is still a fantasy of participation and registration. People are 'accustomed to putting their thoughts online but also, in so doing, they believe their thoughts and ideas are registering' (Dean 2009: 31).

In communicative capitalism the gadget, together with its various integrated social media platforms, provides a foil onto which we can project an illusion that what we say matters and has an effect. By putting communication into circulation, and encouraging the belief that by doing this we are doing something significant, the gadget becomes a 'fetish object', and as such is 'active in our stead' (31). This is a scenario that Slavoj Žižek calls 'interpassivity', and we can understand it as the danger of the fetishism of gadgets. For many users, Dean believes, 'new media let them feel as if they are making a contribution, let them deny the larger lack of left solidarity' (36). She suggests that such fetishes have three 'primary modes of operation: condensation, displacement, and denial' (36). Condensation operates in the reduction of complexity to one thing; displacement is the idea that everyday actions – tweeting, consumption choices and so on – are politically significant, which 'displaces political energy from the hard work of organising and struggle' (36); denial entails rejecting the background and context of technology in order to focus on technological objects as autonomous magical cures for our ills. The danger with such 'clicktivism' is that it allows one to imagine one has an orientation to gadgets as things, when this whole perception is mistaken and any hope of such an orientation is nullified. For example, our relationship to mobile phones is defined by Dean in terms of 'a marketing relationship to oneself; targeted advertisements that urge consumers to differentiate and specify themselves', such that 'systemic problems ... are treated as the effects of individual choices' (2016: 63). Movements and transactions are tracked by these 'technologies and practices of commanded individuality' (63).

Dean refers to Napster as an example of such fetishism; while there was a lot of store put into its capacity to disrupt the commodity form of information, and as such the broader configuration of communicative capitalism, Dean argues that such a view overlooks the wider formations of capital, suggesting that 'Sharing at one level (files) allows ownership at another (hardware, network access)', with the result that 'The technological fetish covers over and sustains a lack on the part of the subject' (2009: 37).

The notion of reflexivity plays a significant role in Dean's critique – she argues that in most areas, from science to ethics to production, reflexivity has become the mode of action; indeed, that 'Techno enthusiasts write as if reflexivity were the solution' (2010: 14). Yet in the absence of a master signifier there is no guiding principle, and in fact 'reflexivity cannot determine for us what we ought to do' because we have a 'reflexivity that goes all the way down'; as such this is 'another name for the decline in symbolic efficiency' (11). While reflexivity is generally understood as something that enriches, here it is presented as a trap – one that has an even more detrimental effect on the subject because it removes any ground, as 'the endless loop of reflexivity becomes the very form of capture and absorption' (13). It has an even more compelling and damaging aspect in relation to the formation of the subject, in what Dean calls the 'reflexive circuit of drive' (38). This is the compulsion to push on endlessly through the morass of information online, the repetition of the same actions and circuits that are never satisfied, but in which we are driven to continue in so much as we find pleasure, or *jouissance*, in failure.

This contributes to a form of subjectivity Dean calls, after Giorgio Agamben, 'whatever being'. The constant circulation of information that is never received or understood – in which content ceases to matter and only contribution counts – produces an ironic indifference towards everything, a 'whatever' response: 'The sender's message ... is neither accepted nor rejected. Rather, the "whatever" response distils the message into the simple fact of utterance' (68). Such a whatever response means we lose the capacity for reflection, while at the same time being caught in a cycle of reflexivity. Reflection is associated with a subjectivity forged in the individuated era of liberal democracy, but it breaks down with the loss of symbolic efficiency and the coming into being of what Deleuze and Guattari call 'dividuals', understood as the kind of subjectivities associated with control societies: 'fluid, hybrid, and mobile subjectivities who are undisciplined, who have not internalised specific norms and constraints, and who can now only be controlled' (75). Thus Agamben 'affiliates whatever being with the capitalist commodification of the human body and technologization of its image in the spectacle' (80).

While Agamben sees political potential in whatever being, Dean is clear in her pessimism that this vision is a passive dead end: 'I can locate here neither a politics I admire nor any sort of struggle at all' (83). The failures of whatever being even extend to arguments that might otherwise redeem it, for example the idea that the conscious breakdown of work time and leisure time could be a practice of liberation. In contradistinction, Dean references the traditions of the libertarian left as manifest in the Paris Commune, a test case of the left imagination for a different kind of collective politics, wherein the fluidity and creativity of a liberated subjectivity is realised in a ferment of political upheaval. Yet this seems even more inconceivable in our society than it was in nineteenth-century Paris. In a society dominated by digital devices, Dean points out, 'an erasure of boundaries looks more like capitalist real subsumption than it does the revolutionary praxis of the oppressed' (2016: 138).

It is therefore difficult to see how a political economy rooted in circulating drive, amongst whatever beings, could offer anything other than submission to the logic of the gadget as fetish object. Caught up in its reflexive loop, it goes nowhere in perpetuity and returns us to a situation where the only relationship to gadgets that could be progressive or helpful would be to set them aside.

Yet, like the other two critical positions discussed, Dean does have a positive response, which is to recognise the power of antagonistic approaches, of the solidarity of groups and the potential for commonality in struggle. She pursues this by arguing for a return to the party form as an escape from the current impasse. Dean's vision of the party is distinct from the party of bourgeois democracy or the Leninist vanguard; rather she proposes it as an opening through which 'the people' can emerge and make themselves heard. The party is the necessary counterpoint to the prevalence of the crowd in the uprisings of the early twenty-first century. 'The crowds breach of the predictable and given creates the possibility that a political subject might appear. The party steps into that breach and fights to keep it open for the people' (5). The crowd impacts on those involved, forcing them into a situation beyond themselves, or, as Dean puts it, '[t]he crowd is more than an aggregate of individuals. It is individuals changed through the torsion of their aggregation' (9). The crowd's impulse needs to be sustained in order that these subjectivities are not captured in the form of the individual, and it is here that '[t]he party knots together unconscious processes across a differential field to enable a communist political subjectivity' (28).

Given Dean's arguments about communicative capitalism, it is difficult to see the crowd's transformation and emergence as a party as involving anything other than a bodily face-to-face process, which is a shame given the potential for the 'digital party'. Yet Dean does refer to a seeming contradiction within the technologies of 'commanded individualism', wherein aspects of technology provide possibilities to escape the command to 'be yourself'. As she puts it: 'the technologies that further individuation – smartphone, tablet, Facebook, Instagram, Tumblr – provide at the same time an escape from and alternative to individuation: connection to others, collectivity' (64). This observation is not fully explained or accounted for, but it does suggest a way in which Dean's emancipatory proposals might be articulated to help us break free of the gadget-object (which I will develop later in this chapter).

These arguments offer powerful critiques that we can apply to the gadget-object, but they also provide us with three positive threads, or what we might call negations of negations. These are *care*, *becoming* and *collectivity*. The capacity which ties these hopes together is that of will, or rather, intention.

GADGETS, WILL AND INTENTION

As a counterpoint to the rather bleak reading of the gadget as techno-capitalist dispositif, intention offers a framework for agency and emancipation, for the realisation of care, becoming and collectivity. It is as well therefore to wind back a little and reflect on this more broadly to clarify the scope of intention in this context.

One of the most helpful frameworks regarding technology's relation to human intention and will is that of cultural materialism, as originally conceived by Raymond Williams. At the heart of cultural materialism, as Roger Silverstone argues, is 'a fundamental belief in the effectiveness of human agency: our capacity to disturb, disrupt and to distract the otherwise cold logic of history and the one-dimensionality of technology' (2003: xi). To some this may seem contradictory, given the general understanding of materialism, at least in its 'orthodox' Marxist formulation, as a calculative mode of thought that reduces history and human agency to deterministic effects of the iron laws of history. Against such perspectives, Silverstone points out, Williams held that 'Technologies may constrain but they do not determine' (xi).

While such a mechanistic understanding of history is often assigned to the tradition of historical materialism, and even more so to dialectical materialism (as framed by the Stalinist-era 'diamat'), these tendencies are by no means inherent in Marxism. It is within this tradition that Williams worked, and out of which his understanding of cultural materialism sprang. His reasons for rejecting a crude determinism are famously explored in his essay, 'Base and Superstructure in Marxist Theory', in which he argues against the notion that culture is merely part of a superstructure, a kind of epiphenomena of the productive base. Early in that essay Williams tells us this conception emerged in the transition from 'Marx to Marxism', and that 'Marx's own proposition explicitly denies this and puts the origins of determination in men's own activities' (1973: 4). Thus, in the context of Marxist theory, the notion of determination itself becomes not one of a prefiguration of activities but one of a 'setting of limits and the exertion of pressure' (6). The integrated nature of culture then becomes key for Williams, such that real concrete activities need to be seen in direct sets of relations 'containing fundamental contradictions and variations and therefore always in a state of dynamic process' (6). This is a position he develops in *The Long Revolution*, where he tells us that: 'I would then define the theory of culture as the study of relationships between elements in a whole way of life' (2011: 67).

For Williams agency and intention are inextricable elements of change and cultural evolution, even if this aspect is downplayed or outright abandoned in other forms of materialism. The perspective of cultural materialism is followed through in Williams' exploration of the relationship of technology with the social, and it is mapped out most clearly in his chapter 'The Technology and the Society' in *Television and Cultural Form*. Here the conception of determination is transposed directly from the 'base-superstructure' relationship onto technology itself.

Williams sets out a number of nuanced variations of the positions he detects in common arguments regarding the effects, specifically, of television, and the idea that 'It is often said that television has altered our world' (2003: 1). While he sets out to explore such notions of the effects of television in particular, these positions are readily generalisable to culture and technological development as a whole. In 'The Technology and the Society' he explores a number of possible assumptions that support the misunderstandings of the influence of technology that he sees as rife. These are broadly divided into the two categories of 'technological determinism' and 'symptomatic technology'. The first defines a set of assumptions that see technology (in this case television) as directly altering our world; in the second, pre-existing social circumstances latch on to new technologies and develop and distribute them. In both these cases technology is understood to be developed in a social vacuum, 'invented as a result of scientific and technical research', 'discovered as a possibility of scientific and technical research', or to have become 'available as a result of scientific or technical research' (3–4). As such:

> In technological determinism, research and development have been assumed as self-generating. The new technologies are invented as it were in an independent sphere, and then create new societies or new human conditions. The view of symptomatic technology, similarly, assumes that research and development are self-generating, but in a more marginal way. What is discovered in the margin is then taken up and used. (6)

Because these positions are so engrained, our thought is diminished and unable to progress beyond understanding technology as a self-acting, isolated and autonomous entity. Williams' primary objection is that such a view dislocates technology from its social context, and this very much applies to gadgets. It is not only that this is undesirable as an essentially bourgeois understanding, but also that it is inaccurate. In these deterministic models the vital element of intention is left out of the picture. Intention should therefore be a central part of any interpretation as a way of avoiding both the deterministic and symptomatic versions of 'self-acting' technology.

In the case of technological determinism, intention needs to be accounted for in order to recognise that technology is developed and used with a purpose. This is of particular importance in the case of technologies of domination or exploitation. Without this recognition, any kind of reformist agenda would not be able to summon a rationale beyond critique alone; that is, challenge the status quo and explore the possibility of developing alternative technologies. Such a possibility of conscious or willed technological change is exactly what a bourgeois conception of technology would be keen to undermine. As Williams argues: 'It is an apparently sophisticated technological determinism which has the significant effect of indicating a social and cultural determinism: a determinism, that is to say, which ratifies the society and culture we now have, and especially its most powerful internal directions' (130). Likewise, the insertion of intention into the symptomatic technology framework would mean that social needs would directly inform technological development rather than existing on its margins. It is this inclusion of intention, as part of a broader complex of technological, social, political and economic factors, that completes a call to understand technology as part of a 'whole way of life', and that demarcates this approach within the borders of a 'cultural materialism'.

This is also a deeply democratic vision and one that applies readily to gadgets as previously defined. There is a current phase of thinking that places us on a concerted and almost inevitable path of servitude to gadgets; however, while we certainly need to be mindful of such thinking, we do not need to submit to the temptations of technological melancholia or of a nostalgia for a time when our technology was simpler and being human was more clear-cut. While gadgets are today clearly more advanced in their capacities and ubiquity than the television analysed by Williams, his basic principle remains compelling. We must not separate the technology from the society, or its economic and political entwinements from its vital entwinement with the human brain.

In order to recuperate intention as a useful theoretical tool we must recognise the human brain as its necessary, though not sufficient, material locus. The brain needs to be understood as part of, but not wholly constitutive of, a thoughtful or intention-generating assemblage. Other necessary elements of such assemblages being, for example, bodies, language, communication networks, broader cultural forms and social infrastructure. This focus on the brain as the locus of intention again evokes the Marxian concept of 'species-being'. Nick Dyer-Witheford argues for the continuing value of this concept in the contemporary analysis of power and agency – not in the sense of offering an essentialist concept that excludes all other forms of agency, but in recognising the self-shaping constituency of the human species as a particularly open process. He reminds us of the distinction between species-being and species-life: the former is 'not just existence as a natural, biologically reproductive collectivity' but 'the human power to intentionally alter species-life'; as such it entails 'material capacity, self-consciousness, and collective organisation, all feeding into each other' (2004: 1). Intention is thus conceived as a collective critical consciousness that directs social change and action. Likewise Williams, with his insistence on class intention, also postulates a certain degree of consciousness in the process of both domination and opposition: 'in relation to the full range of human practice at any one time, the dominant mode is a conscious selection and organisation. At least in its fully formed state it is conscious' (1973: 13).

This short diversion into cultural materialism has given us a concept of intention that, if adapted to the current configuration of the gadget-object, and articulated with care, becoming and collectivity, can offer a legitimisation and purpose for gadget consciousness. Since Williams is rather vague about how intention actually and practically comes about, I will supplement his account by returning to the question of recursion and reflexivity, previously touched on in the context of communicative capitalism.

BEYOND THE GADGET-OBJECT

In considering Williams' idea of intention in the context of gadgets we can connect it with recursion and reflexivity. In *Blog Theory* Jodi Dean (2010) suggests that reflexivity, which in the liberal democratic framework is lauded, is more like a bottomless abyss that contributes to 'whatever being', characterised by the decline of symbolic efficiency and augmented by digital communications. Yet if we are to offer a more optimistic take then it is worth examining the relationship between reflection, reflexivity and recursion as one that can actually strengthen agency rather than pacify it. This is of particular importance here as it involves a set of ideas that will be widened and taken forward into the rest of the book.

Dean's argument about reflexivity and the loss of symbolic efficiency doesn't account for the possibility that there can be communications, including relationships to gadgets, that can still generate symbolic efficiency, in the way of communicative action or even basic forms of pragmatic communication, mutual recognition and affect. In challenging reification and/or domineering reterritorialisation, where Dean argues that communications across the entire internet fail to reach their destination, it is more helpful to consider the micro level of specific utterances. In fact, we can see many instances where pre-existing networks coordinate and formulate consensus positions towards action. While it is true that more arbitrary aggregations formed around general circulation, for example in newspaper comment sections or trending topics on social networks, fail, it is not universally true. In more localised networks of regular interlocutors, or specifically directed utterances, reflexivity is not bottomless but stabilises around clusters of consensus and emotional or affective resonances.

We need to be careful not to confuse reflexivity with recursion. Recursion is the process wherein information loops back on itself in order to create stability or reactivity to an environment. It shares some characteristics with reflexivity, which is the internalisation of such recursion to the point of self-direction. A thermostat is often given as the most basic example of a recursive, or cybernetic, system. Here there is recursion in that the device operates by acting on environmental information which is itself responding to the device's setting; as such, a feedback loop then regulates and stabilises the environment in which the thermostat operates. Language itself operates in recursive constructions. Noam Chomsky argues that recursion is an essential part of natural language, allowing for an infinite number of sentences to be generated from a finite set of rules and resources, with different iterations of words and sentences incorporating elements of themselves. As Steven Pinker explains, 'These rules embed one instance of a symbol inside another instance of the same symbol (here, a sentence inside a sentence), a neat trick – logicians call it "recursion" – for generating an infinite number of structures' (1994: 101).

Recursion is also key for computing and information theory. Tim Jordan defines it as a practice that 'allows a particular process to use either itself or products or elements of itself back within that same process' (2015: 33). Computing develops a more complex pattern of recursion than a thermostat, but nevertheless relies on recursion to function. The original conception of modern computing is commonly understood to be found in Alan Turing's 'Universal Turning Machine', an imaginary device that lays out the principles of a programmable computer, which requires the development of algorithms that incorporate their own outputs as inputs – for example a complex set of calculations such as fractal mathematics. This insight is itself underpinned by the work of the mathematician Kurt Gödel, in that his exploration of mathematical recursion led to the 'first precise definition of an algorithm' as essentially 'a function for which there is a mechanical rule for computing the values of the function from previous values' (Jordan 2015: 37).

More complex systems have a greater capacity to adapt and adjust to variations in feedback, and, in digital systems at even higher levels of development, to alter their own algorithms because of recursion. For example, the Google search algorithm is continuously adapting itself in reaction to each new search and, according to Matteo Pasquinelli (2009), this is largely where Google's value resides. Here, according to Jordan, we see a situation where '"Eating" its own products allows a recursive programme to both absorb its own information and to alter its own functioning' (2015: 38).

The greater the adaptability of the algorithm the more capacity the system has to move beyond adaptation and develop a degree of autonomy – the ability to *act* upon itself, to create ever more variant forms of recursion. This produces several divergent possibilities for control and exploitation, but also for liberation. In terms of exploitation this echoes the arguments given above, but in the context of recursion it specifically concerns information being drawn from users of various privately owned and profit-oriented systems and platforms, and that information being absorbed and recycled into a much wider informational ecosystem. In this ecosystem algorithms sort and re-use information for the purposes of marketing, advertising or other forms of persuasion or 'nudging', which in turn create huge amounts of value for the owners of these systems. As Jordan argues, variation and difference are the key (2015: 40). Each time new information is added, and when it is correlated with and proliferates throughout a system, new and different outputs (and then subsequently inputs) become possible. This is the case with the Amazon algorithm that matches up people with products they may not otherwise buy. It becomes an issue of exploitation, Jordan argues, because it is an issue of property. When something I input is used by the 'owner' of the system or platform to predict what people like me will be interested in, 'something the user "has", in the information they enter, is "taken" by the recursive programme and served up to the controller of that programme' (40).

However, depending on context, this recuperation and control need not be so. In an open source, cooperative or otherwise 'common' system, the 'taken' information can be re-used, reformatted and shared for any number of reasons with any number of algorithms applied. For example, there have been experimental platforms developed that work to aggregate information from multiple sources, bundle them and organise them for the express purpose of informing all users of the outcomes, thereby creating a powerful tool to support active decision making. Two examples of this, that I discuss in more detail in Chapter 5, are the 'X-net' platform for political organising in Spain, and the 'Suki' project that was undertaken by activists and hackers in the 2011 anti-fees protests in London. These took advantage of recursive information, and in each case this was done not to heighten power or accumulate wealth but to share benefit. This is a case where controllers still instigate the process of design, and begin the process of recursion, but there is an active decision on their part to open the process to consultation and pooling of informa-tion. In such a situation great care is required to make sure power and or capital does not accumulate in the hands of an unanswerable individual or elite, given Jordan's point that 'recursion controllers have a considera-ble advantage over those who provide the necessary information to start the process and keep it going, because the controllers can form a stream of new information from recursions' (42). The issue then becomes the relation of the controllers to the users and the technology itself.

The capacity for activists to balance power also carries the potential for significant shared benefit, partly in their being able to collectively manage the results of recursions. The shift from recursion to reflexivity – what one might call a levelling up, or more technically a phase shift – develops this capacity towards taking collective control of recursion. Taking control of recursion means altering the feedback through will and volition, rather than mere difference and variation, whether for an individual or group (as such this becomes an act of liberation). The question is then the degree to which different forms and interventions in recursion might direct or restrict the capacity to balance power. The logic here also leads to the conclusion that it is possible to have exploita-tion (transforming information into a commodity form) while still achieving a certain degree of reflexivity and autonomy. The underpin-ning assumption, and this is one that will be developed and defended in subsequent chapters, is that reflexivity exists where the intensity of the recursion and the degree of autonomy produced moves beyond a system acting on itself to a state in which the system can alter its own initial conditions. As far as current technology allows, such a change of purpose – a capacity akin to volition or will – will always include human actors somewhere in the chain to achieve such collective capacity (this is because of the recursive or 'loopy' nature of the human brain and consciousness which translates recursion into reflexivity – as will be further elaborated in the next chapter).

If we accept Jodi Dean's argument that the (impossible) satisfaction of drive is the force that motivates communicative capitalism, and the impulse to consume is the prescribed route to satisfaction, then the recursive state is one that moves between the inevitable impulse to consume and dissatisfaction (the dissatisfaction fuelling the next impulse). If there is a circuit that entails capture in a credit and debt cycle, then persons in that situation can be understood as caught in a recursive loop – just as Dean describes. However, reflexivity offers the possibility of escaping that closed loop and introducing new starting conditions, or at least working on a change to the regulatory algorithm. It does this when we consider the definition of reflexivity as a mode of being that accounts for itself in terms of recursion (input-output-input), but that also invokes *meaning*, in so far as the system contains itself sufficiently to have a contextual understanding of itself and others around it, and to make decisions based on what is best – not just for itself but also for others. As soon as the capacity for meaning is understood then the argument presented by Honneth, regarding the importance of recognition, can be reintroduced.

Reflexivity entails recursion but recursion doesn't equal reflexivity. In her analysis, Dean interprets reflexivity in a very particular way, which excludes the phase-shifted reflexive/reflective mode (thought), wherein meaning and purpose can *only* emerge in that context. Recursion all the way down does not mean reflexivity all the way down; we still run into material reality at some point: symbols are redeemed in recognition and action, code is executed, bodies are affected. Most significantly we run into the wiring of the brain.

Reflexivity requires a certain intensity and velocity of recursion to take flight, and it is also necessary for – though not equal to – self-consciousness. The significance of this idea is pivotal to the whole nexus of gadget consciousness, as will become clear as we move to look at the nature of the brain and consciousness in the next chapter. It is my contention that, at the level of gadgets as things, these recursions capture the fourfold as present in the thinging of the gadget. This amounts to a capacity for reflexivity and intention – and it is intention that is the key to a progressive and non-deterministic relationship to gadgets.

For the sake of clarity, at this point it is well worth reflecting on how all the elements of the argument I have been developing tie together. I have argued that gadgets, as defined in the previous chapter, operate as part of the dispositif of techno-capitalism. In this dispositif they are pushed towards the form of the gadget-object, manifest as pure commodity, indeed the commodity par excellence of the current era. They are devices that, in combination with the digital networks that constitute the techno-capitalist dispositif, operate to exploit and control.

I suggested critical perspectives to challenge the legitimacy and inevitability of this configuration of the gadget. The first perspective was that of Axel Honneth, who sees the most important problem as reification. The second was that of Félix Guattari, feeding into autonomous Marxism, where the problem is seen as one of the control and territorialisation of the subject. The final critical point of view raised was that of psychoanalysis, for which the pressing issue is our capture by the circuits of drive and the 'whatever being' this induces.

I then proposed cracks or openings in and/or though these frameworks that open up a positive possibility for escape, or at least an alternative take. In the first case I suggested a response to alienation by recourse to the ideal of recognition and the commitment to care; in the second case by exploring the move from control to escape and deterritorialisation in becoming; and the final case by a break out from the recursive circuits of drive with the intensive realisation of reflexivity and collectivity. I argued that all three concerns can be addressed by recovering the idea of intention.

The conclusion is that there is an understanding of gadgets that sits well with the capacity for intention. These arguments have been made to show that a materialist reading of gadgets, as operating on an axis of gadget-objects to gadgets-things, is conceivable within the current configuration of techno-capitalism; that despite the orientation towards the gadget-object we can develop a critical theory that valorises the power of a collective will, providing the initial content for a conceptualisation of the gadget as a thing that things. In that regard we can start to imagine the gadget-thing as operating within a different kind of dispositif and with a different logic. This is a step towards a conception of gadget consciousness as counter-hegemonic; of an active consciousness in the sense of self-consciousness; of consciousness-raising and emancipatory mindfulness, as opposed to the passive descriptive consciousness of hegemony and control.

In the next chapter it will therefore be necessary to take a step back to provide more supporting arguments for the material underpinning of intention. This will mean exploring the place of the brain and consciousness as the nexus of intention, and the context of collective intention in particular.

3
Gadget Brain

In the previous chapters I have explored the nature of the gadget and its material and economic context. I identified the need to understand the relationship between intention and gadgets in terms of consciousness and the human brain. It is therefore to these two focal points that I will turn in this chapter.

Consciousness is famously defined by the philosopher of mind, David J. Chalmers, as the 'hard problem'. As he explains: 'Consciousness poses the most baffling problems in the science of the mind' (Chalmers: n.d.). Indeed, before even defining consciousness, we need to ask what kind of consciousness needs to be defined, so that we can raise the right, or even a meaningful, question.

The most commonly understood sense of the term refers to the first-hand conscious experience of having an experience. As Thomas Nagel (1974) frames it, there is something that it is like to have an experience; and as Chalmers says, 'The really hard problem of consciousness is the problem of *experience*.'

This is consciousness as a mental state, and it brings up a swathe of problems pertaining to, amongst other things, the relationship of mind and matter (or mind and brain). Are experiences something that can be reduced to the material substrates of the brain, understood to be the locus of thoughts? Or are they not associated with the material world at all, but with something of another order? If so how are the two orders related? The issues multiply very quickly, to include the question of to whom experiences belong, and indeed if they need to belong to anyone. Thus the question arises of what it is to be or to have a self. Is the self a real thing? Furthermore, do experiences need to be experiences of something, or can they be purely internal, in the way of thoughts, reflection or dreams. Can one have experiences of which one is not actively conscious, but which somehow become part of consciousness?

Once we get beyond the initial questions of experience, awareness, ownership and so on, there is then a further set of problems concerning the extent to which consciousness is active or passive. Do conscious minds actively 'aim' at the world and construct a picture of an outside reality, in the mode of what is known as 'intentionality' (as distinct from intention), or do they passively absorb the world? Indeed, how much are minds even aware of what they are absorbing? Another very significant issue, certainly in terms of the approach in this book, is where the borderline of consciousness sits. Does consciousness begin and end with the confines of the body and brain, as limited to a single human organism, or is it a more extended *activity* that incudes tools, environments and other beings?

Given that this is not a book on philosophy of mind or neuroscience, and what we are confronting is, after all, the 'hard problem', it will be impossible to give all these concerns and questions a full hearing. This chapter will therefore aim to support a working definition that consciousness is a materially based phenomenon that entails the capacity to reflect and act and to make decisions, whether individually or collectively. This is a definition that allows us to ascribe will, or rather intention, to consciousness, and therefore to address its political and ethical significance. These are characteristics normally ascribed to a self, which will also be under consideration here. The chapter will end by making an opening case for extending the idea of consciousness to include our gadgets, and argue that this combination allows for a shared consciousness with elements of collective will and action. For that purpose, it will be necessary first to set out an argument as to what the scope of consciousness is, and then step back to consider how this relates to the material world, the brain and gadgets.

Much philosophy of mind and political thought sets out from the assumption of singular minds and separated individuals, and thus places the onus on proving a path from the individual to the multiple or the collective. Yet the idea that consciousness is primarily an individual property of a single mind/brain/body overlooks so much of its supporting mechanisms and social forms that it is ontologically questionable at its foundations.

The notion of a bounded and limited consciousness is a product of a specifically individualistic and bourgeois world-view. Christopher Caudwell, though not a well-known figure today, raised this concern when he understood the science of psychology to be in a state of perpetual confusion. This is because of the ideological need for the bourgeois individual to appear 'free "in himself"' and 'standing in a domineering relation to his environment', to be 'determining society and not determined by it' (1977: 135).

Such a belief needs to be held in balance with the widely held belief in the superiority of the positive sciences, in which the notion of a unified and deterministic universe holds sway. What this identifies is the temptation to focus on the brain as a unique and discrete case, and thus to see the consciousness problem as one that is defined by an existing set of assumptions, in particular that we are autonomous beings with a primarily determining relation to nature and society. As Caudwell points out, this is an illusion, one he believes is perpetuated by the bourgeois 'domineering relation to capital and ownership of social labour power' (135). In contrast, he argues that 'The sum of bourgeois wills produce history, but it is not the history any one bourgeois willed', and the bourgeois themselves failed to recognise that 'all the phenomena that constitute the Universe are mutually determined' (135). In that regard I believe we should start with the broader question of consciousness in general.

Such a starting point puts us in mind of the notion of absolute consciousness, which pre-dates the era of neuroscience and psychology. Absolute consciousness, that is consciousness as all-encompassing totality – reality in itself – is the end point of the thought of G.W.F. Hegel. While I will take a different view from Hegel, one that focuses on the interrelation of multiple minds, his argument is so pivotal in the history of philosophy that it needs to be taken into account. According to Hegel, individual consciousness is always already relational; that is, mutually attributed between subjects: 'Self-consciousness exists in and for itself when, and by the fact that, it so exists for another; that is, it exists only in being acknowledged' (1977: 111). This dialectical understanding renders interlocutors as simultaneously self-conscious entities who 'recognise themselves as mutually recognising one another' (112). But there is a tension here between self-consciousness and the object of consciousness. There is a movement between the positing of the self, its necessary apprehension of objects, and its reliance on the recognition of others – each element is separate but dependent on the others, and each struggle to reconcile this contradiction. The movement of the contradiction is towards a supersession of the object and the other, towards an understanding and knowledge of the whole situation rather than a partial point of view. This overarching knowledge develops towards what Hegel refers to as 'absolute knowledge', which is not the omniscient point of view of a god, but the coming to awareness of the place of the human mind in self-consciousness. Such absolute idealism is distinct from a subjective idealism, which would reduce reality to a multiplicity of individual irreconcilable realities. Peter Singer captures this distinction when he explains that Hegel 'calls his form of idealism *absolute idealism* to distinguish it from subjective idealism. For Hegel there is only one reality, because, ultimately, there is only one mind' (1983: 72).

Hegel thus conceives reality to be entangled with the collective mind or 'spirit'. It is in the grasping and structuring of experience that reality itself takes shape, and in the dialectic of self-consciousness that humans grasp it. This entails a specific perspective on the commonality of mind in which mind is the common ground of the human, and of reality as such. While individual humans have distinct perspectives, they are all part of the singular mind that underpins reality. As the coming to awareness of the place of mind, absolute knowledge is not knowing about everything, an accumulation of the totality of facts, in the way of a crude empiricism. As Singer puts it: 'Reality is constituted by mind', or as Hegel states, it is 'mind knowing itself in the shape of mind' (1983: 71).

The early twentieth-century Hegel scholar G.W. Cunningham makes the case that 'whatever Hegel calls "absolute knowledge" is simply the result of his consideration of thought as it appears in every knowing experience' (1908: 620), and that 'Hegel bases this conception of absolute knowledge directly and unequivocally upon our common knowing experience' (621). This is the fulfilling of the interactions between knowing subjects within and through the dialectic. The absolute has its seeds in the earliest most basic kinds of sensual experience, but only comes to fruition via 'a careful and painstaking examination of all stages of consciousness' (624). Thus, it is specific acts of knowledge which are oriented towards totality through commonality. And it is the totality of absolute knowledge that is the end of the dialectic.

According to Alexandre Kojève, Hegel is demonstrating 'what the Man must be who is endowed with a Knowledge that permits him completely and adequately to reveal the *totality* of existing Being' (1969: 31). This is possible because a person, located at a certain point in history, becomes a microcosm of the moment, and the knowledge leading to that moment. As such, Hegel himself 'was a microcosm, who incorporated in *his* particular being the completed *totality* of the spatial-temporal realisation of *universal* being' (35). This entails the reflexive moment in which he, Hegel, 'was able to understand the *World* in which he lived and to understand *himself* as living in and understanding this World' (35). Desire lies at the root of this, the moment of contemplation of something in which the contemplator gets lost, only to be interrupted and be made aware of themselves through some internal desire, be it hunger, desire for comfort or some other interruption, that produces *self-consciousness*, that is, the split between subject and object and the supersession of the latter by the former. This dialectical movement is the tension that unfolds, inevitably, towards absolute knowledge.

Hegel offers us an understanding of human consciousness as being far more than simply the apprehension or awareness of an environment or a determinate set of facts; rather, it is something that contains within it the collective history and totality of being. It does not begin or end at the confines of the body or brain, nor is it some ineffable religious spirit that operates in an unreachable or inexplicable realm. However, without going into the nuanced debates in Hegelian scholarship, for the purposes of the argument here we can raise a serious question about the 'absolute' in absolute knowledge, in the way it is framed by Hegel. Hegel provides the ground for what we think of as collective consciousness, but there is a serious question as to how helpful that is. While the absolute gives us a window into a realm of consciousness that needs to be recognised, in many ways it overreaches and is clearly of its time. Its totalising conception of reality also raises major ethical and political concerns, in particular the elimination of difference in the absolute idea. It poses the danger that a particular historic state might be fetishised as the final and perfected state; it also leads to the rather overburdened claim that one person can come to embody the completion of world history, which Hegel rather grandly saw himself as having done.

Hegel's totalising vision based on an idealist ontology was famously set 'on its head' by Marx's materialist inversion. As was discussed in the previous chapter, this book is taking an avowedly materialist approach, so whatever the merits or failings of idealism it will be set aside here. Yet we don't want to lose all its insights or its more expansive mode of thinking compared to the narrow bourgeois or positivist framework of the kind Caudwell identifies. Hegel does offer up vital insights that we need to take forward – the sociality of consciousness, the contradictions of thought and experience as drivers of history and change, the collective character of knowledge and the human capacity for self-knowledge and development – yet these must now be set alongside the lessons of mate-rialism and the leaps forward in the neurological sciences.

In his critical exploration of Hegel, the Marxist thinker and cultural theorist Henri Lefebvre presents a way of rescuing some of Hegel's insights while setting aside his grander idealist claims. Lefebvre argues that in Hegel the development of mind towards the absolute idea is still primarily located within existing human minds, and that it is highly questionable as to whether 'the limited mind of an individual, of a philosopher, should be able to grasp the entire content of human experience' (1968: 48). Rather, he argues that 'the content will be attained only through the joint efforts of many thinking individuals, in a progressive expansion of consciousness. Hegel's own claim encloses and limits the content and makes it unworthy of Mind' (48). This simple but profound claim is a root idea for the kind of consciousness I believe gadget consciousness to be. It involves the integration and contestation of the multiplicity of minds, but with the potential to become even more than what Lefebvre describes in terms of the capacity to think and act together, given the affordances offered by gadgets and our potential to relate to them as things.

For this reason, and for the purposes of the present discussion, I will set the Hegelian idealist paradigm aside as being too expansive, too totalising, and too wrapped up in absolutes. Rather, this chapter will argue that shared consciousness can be achieved in a materialist framework within which the working hypothesis must be that consciousness is enabled and supported by the neural substrates of the brain, but is not limited by that substrate and is not located *only* there. For example, the human species shares a great deal through language, culture and affective interactions, all of which are necessary for the form of self-consciousness that defines the species-life of the human, and that inaugurates the capacity for will, intention and self-development. Before developing this point, however, these claims need to be undergirded by an understanding of the relationship of consciousness with the material substrate of the brain.

THE MATERIAL SUBSTRATE:
BRAINS, CONNECTIVITY AND EMOTION

Brains think. The general consensus derived from empirical evidence is that the brain is the material substrate of thought and consciousness. All current credible approaches in neuroscience and cognitive science find their common root, and a significant set of common features, in the theory of connectionism. According to connectionism, there is no identifiable element of the brain that in and of itself explains consciousness. The basic building block of the brain is the neuron. The role of neurons is to 'convey signals from one place to another', and they 'tend to resemble trees, with branching roots to receive incoming messages, a smooth trunk to transmit them and a bushy crown through which they send their messages on' (Zeman 2002: 42). There are around 86 billion neurons in the human brain. When an individual neuron is examined under a microscope there is nothing exceptional about it; it does not differ in substantive ways from neurons in the brains of other animals, even insects. The human brain's only exceptional and unique feature is the size of its operation: the vast number of neurons and the even greater scale of their connections – each of the 86 billion neurons has around 1,000 synapses connecting it to other neurons. It was the early twentieth-century biologist Santiago Ramon y Cajal who identified neurons as the cells that compose the brain, and observed that 'our abilities depend on the way neurons are connected, not on any special features of the cells themselves' (Bor et al. 2017).

The size of the human brain is a result of evolution, the evidence of which is embedded in a number of its features and capacities. The oldest elements of the brain are in the hindbrain, which includes the medulla oblongata, pons and cerebellum. These regulate automatic behaviour such as breathing and facial expression, as well as aspects of bodily movement, and their workings are buried far beneath any conscious awareness. The same is true of the midbrain, which oversees certain physical actions and the release of neurotransmitters that act as the communicating link between synapses. These include dopamine, which is pivotal to reward systems and vital in learning and also in compulsive behaviour. Moving into the forebrain we find more complex and flexible functions; for example, the thalamus directs sensory information into the cerebral cortex, and the hypothalamus, amygdala and hippocampus contribute towards emotions and the formation of memories, and the control of intentional movements that are initiated in the cerebral cortex.

The development of the cerebral cortex, the latest part of the brain to evolve and which sits as its outer layer, is widely understood to be what makes consciousness possible. It is astonishing in its complexity, with around 10 billion internal connections, but only around 1 million neurons feeding information from the rest of the brain. The cerebral cortex then operates with a relatively low bandwidth of input, but massive internal connectivity. It is this connectivity that underpins consciousness and thought, creativity and intention.

Consciousness arises in the cerebral cortex, but in conversation with the rest of the brain. The biological evolutionary explanation is that consciousness provides a systemic overview: while all the autonomic systems operate independently, the cerebral cortex interacts across the whole brain with all these other systems to synthesise a whole organism perspective. As the neuroscientist David Eagleman explains, 'consciousness is the system that has this unique vantage point, one that no other subsystem of the brain has. And for this reason, it can play the role of arbiter of the billions of interacting elements, subsistent subsystems and burnt-in processes. It can make plans and set goals for the system as a whole' (2015: 99). And as two other notable neuroscientists, Gerard Edelman and Giulio Tononi, argue, 'conscious experience appears to be associated with neural activity that is distributed simultaneously across neuronal groups in many different regions of the brain' (2000: 36). This is in line with the idea that consciousness is an emergent entity. That is, it does not exist as a specific thing, or particular quality of the brain, but emerges from the multiplicity of activities acting in concert. The result is that a state of resonance exists when sufficient cross-brain coordination and consilience occur. In the most direct terms: 'Although the theoretical details are not yet worked out, mind seems to emerge from the interaction of the billions of pieces and parts of the brain' (Eagleman 2015: 214). It seems that consciousness only emerges when the brain lights up as a whole, when its small world network is fully activated and interacting.

Emergence does not mean that consciousness is simply an epiphenomenon, rising from and situated above the mechanics of cellular action; it is fully part of the human brain, feeding back into the very fabric of its operation. As Zoltan Torey has argued, 'consciousness is not some newly acquired "quality," "cosmic principle," "circuitry arrangement," or, "epiphenomenon," but an indispensable working component of the living system's manner of functioning' (2014: 8). The conscious aspect of the brain acts to coordinate and integrate different systems, which also sit behind our self-reflexive awareness and capacity to capture and redirect attention and actions. Torey refers to this representation as an 'endogram': a representation of the body's internal states in a centralised clearing house that is the brain, that has evolved to coordinate the incoming signals from various senses. It is a '[c]ontinuous internal representation ... a multimodal situation report' (16).

It is when the consciousness of the endogram is included in the endogram itself that we have a phase shift, at which point we achieve reflexive self-awareness: 'the brain's handling of itself and the reflexive awareness of what it was doing as it was doing it' (20). This also entails an extra strand of what we ordinarily call volition, or will, and what Torey refers to as cosalience: 'cosalience is the sense of self or agency that we feel whenever we speak or think' (24). Language plays a crucial role here, opening up huge affordances out of the awareness that self-consciousness brings, including decision processes that go 'off line', i.e., that can be planned for and abstracted from the present moment. In this way, the 'language equipped brain becomes the source of its own causal leverage' (25).

Antonio Damasio, the noted philosopher and cognitive scientist, agrees that consciousness rests on a neural substrate, but makes the case that a large component of this is actually based in emotion. He argues that there is a tripartite development of consciousness and self: 'the protoself and its primordial feelings; the action-driven core self; and finally the autobiographical self, which incorporates social and spiritual dimensions' (2010: 10). The protoself is the automatic support system which underpins the 'higher' stages of consciousness; it is 'an interconnected and temporarily coherent collection of neural patterns which represent the state of the organism', and as such 'we are not conscious of the proto-self' (Damasio 2000: 174). The core self is the opening of self-awareness, but offers up only a transient awareness of the present; it is not characterised by linguistic understanding but is born when 'our organisms internally construct and internally exhibit a specific kind of wordless knowledge' (168). The core self is activated in our encounters with objects, where there is an interaction of body and object followed by a bodily reaction and a new bodily condition. This forms a narrative, and a knowledge of self emerges in 'the feeling of knowing' (169). So it is that the 'core self inheres in the second order nonverbal account that occurs whenever an object modifies the proto-self' (174). Thus we are conscious of modifications to our protoself, and this becomes a continuous feeling of knowing: there is a '*felt* core self' and 'you know you exist because the narrative exhibits you in the act of knowing' (172).

This then shifts up to the level of the autobiographical self, wherein our high-capacity memory is able to store and categorise vast amounts of data, constructing a narrative of the past, stitching it into the present – or rather the very near past – and from there projecting it into the future.

'The consequence of that complex learning operation is the development of autobiographical memory' (173), which continually expands and develops throughout our lives. Damasio argues that what is distinct about the autobiographical self is its capacity to summon up a self-concept via off-line access to a lifetime of memories '[b]ased on permanent but dispositional records of core-self experiences. These records can be activated at any time and turned into explicit images'(175).

We cannot have an autobiographical self without a core self and a protoself. 'All three stages are constructed in separate but coordinated brain workspaces' (Damasio 2010: 181), and as such the felt emotional states of the core self are an inextricable part of the self-conscious auto-biographical self, and the conscious self-reflective stages of the latter feed back into the former.

Even Damasio – placing huge emphasis on emotion as he does – holds the view that consciousness is a coordinating mechanism. He argues that consciousness is the process whereby the whole organism comes to be represented within the brain, and that 'the organism, as represented inside its own brain, is a likely biological forerunner for what eventually becomes the elusive sense of self' (2000: 22). These mechanisms, in particular the feedback effect of the autobiographical self into the proto and core selves, provide the capacity to counter or at least prefigure affects, and for intensities to be drawn into a process of rational and volitional steering. Thus, we come to the position of thinking not only of consciousness, but of the way that the self is constituted within consciousness, and the place of this within the greater picture of inter-subjectivity and action.

EGO TUNNELS AND THE BOOTSTRAPPING OF SELF

We have established that consciousness is a process, not a discrete object that can be directly seen or prodded. While it is predominantly produced by the cerebral cortex it still cannot be located only there, or in any specific set of neurons with particular characteristics. We have identified that consciousness exists in relation to the brain as whole and cannot operate fully without its emotional centres, but now we need to reflect on the experience of consciousness, that is, to think about what consciousness means for the self and the will.

Thomas Metzinger describes the self as an 'ego tunnel'. It is rooted in the neural activity of the brain, but it is not to be located there as such. Rather, the ego tunnel is a consistent affordance that is effective in generating an experience of the continuity of a self. It is associated, Metzinger argues, with a 'phenomenal self-model' (PSM), which is 'the conscious model of the organism as a whole that is activated by the brain' (2009: 4), akin to Torey's 'endogram' or Damasio's autobiographical self. The phenomenal self-model is established through the experience of having a body, and is extended through the body, characterised by the sense of ownership and unity that allows for the experience of a bounded and separate self – that is, an ego, a subjective experiencing 'I' – and as such the 'content of the PSM is the Ego' (4).

The ego experiences itself through this phenomenal self-model but it is transparent, in the sense that there is never any experience of the experience of having an experience – we never catch ourselves in the act of being. This accounts for Metzinger's metaphor of the tunnel, wherein we are experiencing an illusion of direct access to reality because the means of experience, our senses and brain activity, filter themselves – as well as much of what is external to us – out of the picture, so they are invisible to us. Hence, he claims that '[o]ur conscious model of reality is a low-dimensional projection of the inconceivably richer physical reality surrounding and sustaining us' (6).

Since we are never aware of what we are not seeing – because it is outside the tunnel – what is inside the tunnel appears to be the totality of reality, and that includes our PSM, as the central entity having the experience. The two elements of the PSM and the experience are bound together in a unified composite. Metzinger thus claims that '[w]henever our brains successfully pursue the ingenious strategy of creating a unified and dynamic inner portrait of reality, we become conscious' (6), and this consciousness constitutes the self-consciousness of the ego. The unique aspect of human consciousness is that it also operates on the meta-level of a consciousness of consciousness, a self-awareness wherein the conscious experience contains a version of itself. This recursive aspect turns into reflexivity, wherein it becomes a part of the picture of the world of which it is a part. As Metzinger puts it, '[w]hat sets human consciousness apart from other biologically evolved phenomena is that it makes a reality appear *within itself*. It creates inwardness; the life process has become aware of itself' (15).

On a slightly more technical level Metzinger argues that the ego tunnel is a 'complex property of the global neural correlate of consciousness (NCC). The NCC is that set of neurofunctional properties in your brain sufficient to bring about a conscious experience' (9). In other words, consciousness is a fully material entity, it is a product of the functioning of the neurons in the brain that model reality and allow us to negotiate the outside world. Conscious experience is not something exclusive to human beings, and Metzinger postulates that many creatures may have tunnels, but without any first-person experience: 'indeed, many of the simple organisms on this planet may have a consciousness tunnel with nobody living in it. Perhaps some of them have only a consciousness "bubble"' (64).

However, for a first-person perspective, self-consciousness – or indeed a self – is vital as the meta-level, the reality within a reality. The dynamic aspect of this is the unity of consciousness, the integration of sense experience and the self-model into a single entity that appears as ego, and that requires the emergence of a unified perspective (the tunnel) from the vast underpinnings of the NCC. Metzinger tells us: 'The global neural correlate of consciousness is like an island emerging from the sea … it is a large set of neural properties underlying consciousness as a whole, underpinning your experiential model of the world in its totality at any given moment' (28).

The temporal aspect of self-consciousness is important because it provides the reality effect, which is vital in terms of evolution and survival, and provides a filter that allows the embedded functions of bodily regulation and habit to be undertaken beneath the surface of consciousness, without fragmenting the unified ego tunnel. As such the NCC can be understood as 'an information cloud hovering above a neurobiological substrate', and the 'cloud is physically realised by widely distributed firing neurons in your head' (29).

As we have established, it is currently believed that the firing of the neurons in the NCC has a pattern of alignment: neurons across distinct areas of the brain synchronise and resonate, wherein consciousness emerges through this complex global integration. Metzinger argues that 'there are some indications that the unified whole appears by virtue of the temporal fine-structure characterising the conscious brain's activity – that is, the rhythmic dance of neuronal discharges and synchronous oscillations' (29). This view is also supported by Wolf Singer, who in an interview with Metzinger postulates that 'more and more evidence has supported the hypothesis that synchronisation of oscillatory activity may be the mechanism for the binding of distributed brain processes' (68). The upshot of this is that conscious experience can be understood as a 'special global property of the overall neural dynamics of your brain, a special form of information-processing based on a globally integrated data format' (29).

The global character of this rising of the island of consciousness has also been named by Bernard Baars as the 'global workspace'. Consciousness taps into the global functioning of the brain, and the more connected, the more resonant and the more synchronised the neurons, the more vivid is the self-conscious experience. As Baars explains, 'Global Workspace Theory is based on the belief that, like the cells of the human body, the detailed workings of the brain are widely distributed. There is no centralised command that tells neurons what to do ... the adaptive networks of the brain are controlled by their own aims and contexts' (1997: ix). Consciousness plays a very significant role in that it accesses and coordinates multiple unconscious aspects of specialised brain function, such as colour or angle perception, that are highly compartmentalised. As such consciousness is 'a facility for *accessing, disseminating,* and *exchanging information,* and for *exercising global coordination and control*' (7).

This is evolutionarily advantageous because it affords a great deal of flexibility to human consciousness; it makes us alert to the unpredictable and the aleatory, as we can never be sure what we might have to deal with next, and which part of the brain will be needed to respond. Metzinger tells us: 'Part of Baars's idea is that you become conscious of something only when you don't know which of the tools in your mental toolbox you'll have to use next' (2009: 56). One of the results of which is that when actions and perceptions become habituated they drop below the horizon of consciousness. Examples such as breathing, walking or natural language use are the most fundamental and ubiquitous hidden actions, and of course there are numerous technical examples such as driving a car or riding a bicycle.

The dividing line between the unconscious and the conscious can thus be correlated to the degree of global connectedness and the extent to which such connectedness contains a model of itself. Metzinger recognises this recursive character in several contexts, but one of the most striking is visual perception, where 'the almost continuous feedback-loops from higher to lower areas create an ongoing cycle, a circular nested flow of information'. Metzinger calls this a 'standing context loop', and such a loop is a necessary aspect of consciousness for the PSM because '[c]onscious information seems to be integrated and unified precisely because the underlying physical process is mapped back onto itself and becomes its own context' (31). As such the brain only becomes autonomous and able to make reflexive, planned and meaningful decisions by virtue of sets of feedback loops of self-awareness – of which intention, symbolic language use and meaning become central elements at higher levels of consciousness and volition.

STRANGE LOOPS

The cognitive scientist and philosopher Douglas Hofstadter describes something similar with his concept of the 'strange loop' (2007), which is the essential element that ratchets up our brains from a maelstrom of neural chatter towards reflective self-consciousness. Hofstadter focuses much more intently on this aspect of consciousness and self than does Metzinger, making the claim that subjectivity and selfhood itself is in effect a strange loop.

Hofstadter builds his picture of the strange loop via a synthesis of Bertrand Russell and Alfred North Whitehead's system of types, outlined in their *Principia Mathematica*, with Gödelian mathematics, neuroscience, cybernetics and the philosophy of causality in emergent systems. It is difficult to do justice to this elaborate theory in a short exegesis, but given that the concept of the strange loop is so helpful to the argument made in this book it is at least necessary to give a headline description of the nature of the strange loop as Hofstadter sees it.

Hofstadter supports his argument by explaining the details of the system outlined in the *Principia Mathematica*, or 'PM' as he refers to it, and even more significantly he explores Kurt Gödel's critique of PM, commonly known as the incompleteness theorem. In PM Russell and Whitehead attempted to establish a solid logical foundation for mathematics in order to get around the problem of the contradictions that emerge particularly in set theory in cases of self-reference. Hofstadter discusses such a contradiction via the question of interesting numbers: what is the first *uninteresting* integer? '1 is interesting because 1 times any number leaves that number unchanged. 2 is interesting because it is the smallest even number' (2007: 106) and so on. However, there is a problem because as soon as we reach a number that does not have any interesting qualities one can say that its being the first uninteresting integer is itself interesting, so the idea of the smallest uninteresting integer 'backfires on itself'. This, Hofstadter explains, 'is the kind of twisting-back of language that turned Bertrand Russell's sensitive stomach' (106).

Russell tried to neutralise such problems by instituting a set of rules concerning what level of description could be applied to any set, in an attempt to fend off such self-referential mishaps and create a complete logical basis for mathematics. The goal of PM was thus to establish that 'no set could ever contain itself, and no sentence could ever turn around and talk about itself' (113). Kurt Gödel fundamentally undermined this logic with his incompleteness theorem. Gödel took the formal notation for the logic that constituted PM and translated it into number form, assigning integers to specific logical functions – what is now called Gödel numbering. His insight was that 'once strings of symbols had been "arithmetized" (given numerical counterparts), then any kind of rule based typographical shunting-around of strings on paper could be perfectly paralleled by some kind of *purely arithmetical calculation*' (133). This method could then be used to insert statements about PM into itself, allowing 'PM to twist around and see itself' and make for such statements as 'This very formula is not provable via the rules of PM', which boils down to the claim 'I am unprovable' (138), thus undoing Russell's attempt to ban self-reference from the formal system. The upshot was that a theorem in PM could be true but not provable, and conversely that false statements could be provable. Gödel thus managed to undermine not only PM but also the claim of any system of logic and/ or mathematics to be complete, with no gaps or flaws.

The special feature upon which this rests is the recursion of theorems – their loopiness. Hofstadter argues that such loops are utterly ubiquitous in nature, as well as in human beings and culture. While many loops are just negative or self-contradictory, there are special cases that are the 'strange' loops. The aspect of strangeness 'comes purely from the way in which a system can seem to "engulf itself" through an unexpected twisting-around, rudely violating what we had taken to be an inviolable hierarchical order' (159). The basis for this is demonstrated by Gödel's number system, wherein there is an isomorphism between two systems (in this case with PM) but the higher order system can loop around and influence the lower order. There is then an analogy with our high-level conscious thoughts, which are grounded in the micro-level operations of individual neurons and the multiple connections of the brain, but which still can have causal feedback into the overall system. Thoughts are emergent entities, and at a certain level they gain a degree of autonomy over the lower level 'causal' physical elements of the system. This autonomy produces what Hofstadter calls 'downward causality'.

The power of thoughts rests on this analogy; that is, on the capacity of one set of symbols to isomorphically represent another set at a higher level of abstraction. It is thanks to the existence, and sophistication, of the symbol system in humans that concepts can '*nest* inside each other hierarchically and such nesting could go to arbitrary degrees' (83). Nesting is what differentiates between simple reception and perception: in the first instance a system merely registers and reflects or reproduces an input; in the second a symbol is triggered, and the clusters of neurons in the brain manifest 'a triggerable physical structure' (75). It is from the perpetual looping and nesting of these symbols that the concept of the I emerges, the recursion of which – intermeshed with the self-perception of a looping system turning its focus on itself – bootstraps itself and locks in the I with a solidity that gives it an apparently unshakable reality.

The strangeness of the I thus rests on two important elements, the first being the human capacity to think, which is itself enabled by words and symbols, so that 'our extensible repertoires of symbols give our brains the power to represent phenomena of unlimited complexity and thus to twist back and to engulf themselves via a strange loop' (203). The second element is our inability to perceive beyond this high-level symbolic representation to what is going on at the micro level in our brains. We don't ascribe a self to video feedback, but we do ascribe one to the feedback loops in our own brain, because that operates at the level of the experience – we do not perceive the brain's neural activity in the way we do the pixilation of the video feedback: '[i]ntellectually knowing that our brains are dense networks of neurons doesn't make us familiar with our brains at that level' (204). It is just as well that it doesn't make us too familiar with our own brains, because it is this blind spot which allows us to, metaphorically speaking, overlook the engine and experience only the motion of the car and the turn of the steering wheel. The upshot of which is that '[t]he most efficient way to think about brains that have symbols – and for most purposes, the *truest* way – is to think that the microstuff inside them is pushed around by ideas and desires, rather than the reverse' (176).

Without symbols we would miss the higher-level picture in which 'the level-shifting acts of perception, abstraction, and categorisation are central, indispensable elements. It is the upward leap from *raw stimuli* to *symbols* that imbues the loop with "strangeness"' (187). The strange loop chimes well with both the notion of the ego tunnel and the global-workspace approach to self-consciousness. Beyond the self alone, the notion of a strange loop offers a powerful concept to understand mutual relations between selves, collectivity and our relationship with things.

THINKING AND FEELING TOGETHER

If consciousness emerges from 'strange loops' then, Hofstadter argues, different strange loops are perfectly capable of becoming enmeshed to share perceptions, thought patterns and much else. Indeed, without these shared loops our worlds would be rather impoverished. Hofstadter discusses the entwinement of strange loops as an integral aspect of self. He again draws on the analogy of video feedback loops, observing that when a camera is pointed purely at its own output then it contains a kind of purity in its infinitude, but that when it catches the edge of something else or incorporates something from outside its own loop then that can get sucked into the loop. The kinds of external elements that get drawn into the loops of self are ever present – interactions with others, photographs, conversations, and so on. Just as the presence and words of others are absorbed into our own loops, so goes our own presence with others. Thus, we mirror others in our brains and we are in turn mirrored in others. We represent, and are represented by, these entwined loops that operate at 'vastly differing levels of detail and fidelity inside our cranium, and since the most important facet of all those human beings is their own sense of self, we inevitably mirror, and thus house, a large number of other strange loops inside our head' (Hofstadter 2007: 207).

Hofstadter offers an example to distinguish what he means by levels of fidelity. He talks movingly about the experience of the loss of his wife, and explains that what provided some comfort to him after her death was the impact of looking at a picture of her and realising that a 'core piece of her had not died at all, but that it lived on very determinedly in my brain' (228). This was the case because knowing someone well enough is to understand their perspective, way of thinking, history and self. Hofstadter tells us that with a marriage, a third entity emerges, and that with his late wife '[s]omething, some *thing*, was coming into being that was made out of both of us ... we were one individual with two bodies, one sole "pairson", one "indivisible dividual"' (223).

This happens because what constitutes a self is not inherent in the material substrate but in the 'very vast *pattern*, a style, a set of things including memories, hopes, dreams, beliefs, loves, reactions to music' (230). As such, what this means is that consciousness is distributed. This is not to say that we are of one mind, or that in a quasi-mystical sense we all share one cosmic soul, but that we share our selves, at least to an extent, between bodies. The dividing line between selves is still powerful, in as much as our loop of self is always primarily located in one brain, given that the loop of self-perception is always necessarily inwardly pointing. Nevertheless, '[a]lthough any individual consciousness is primarily resident in one particular brain, it is also somewhat present in other brains as well, and so, when the central brain is destroyed, tiny fragments of the living individual remain – remain *alive*, that is' (230).

This connection goes beyond reciprocal relations, but is actually an overlapping subjectivity, a kind of *meta*subjectivity. In such a scenario, the 'other' person is partially reproduced and alive in the second brain, so in Hofstadter's case being so close to the 'personal gemma' that was his wife 'had brought into existence a somewhat blurry, coarse grained copy of itself inside my brain, had created a secondary Gödelian swirl inside my brain' (233). Such a 'swirl' will always be coarse and limited, given that there is only ever a secondary incorporation of the other's strange loop, but this is still much stronger than mere intersubjectivity, with its exchange of discourse between interlocutors being just a surface level interaction.

Our brains and thoughts can undertake these incredible acts of flexibility and adaptability because they have features in common with computers; that is, they are 'universal machines of a different sort' (245). Hofstadter crucially sees thoughts as synonymous with consciousness, in as much as thoughts are symbolic, conceptual and self-reflective: 'Consciousness is the dance of symbols inside the cranium. Or, to make it even more pithy, consciousness is *thinking*' (276). Thinking thus includes the capacity to model the world using analogy; just as numbers are isomorphic to the logic of PM, our 'neural hardware can copy arbitrary patterns', and as such, '[t]hrough our senses and then our symbols, we can internalise external phenomena of many sorts' (245). And crucially we have developed even beyond copying patterns, to actively modelling ourselves and other beings. This means going through what he calls the 'Gödel-Turing threshold' – it is the self-modelling which gives us consciousness and an I, and it is the modelling of others which means that we're always already hybrids: 'every normal adult human soul is housed in many brains at varying degrees of fidelity and therefore every human consciousness or "I" lives at once in a collection of different brains, to different extents' (257).

Hofstadter argues that this thought and mutual modelling is entirely bootstrapped though the recursions of the strange loops; there is no hidden perceiver observing the dance of symbols, it is simply symbols upon symbols in circulation operating in ever greater hierarchies of nested concepts. At each level symbols capture the lower levels that support them and as such give an ever greater and more abstract perceptual and causal capacity as the loop intensifies. The concept of the I that is brought about and brought to life is not just some identifiable part of the brain but rather the pattern of levelling up from raw input towards symbolic representation upon symbolic representation, until a point is arrived at when symbols can catch the whole situation. An important consequence of this levelling up is that these abstract symbols – especially the strange loop that is the I – are no longer simply epiphenomena but causal components in the given situation, and in the self. As a result, such ephemeral entities as symbols and strange loops, which are essentially patterns, are able to act on and influence the movements and actions of the components that make them up. The I is an emergent phenomenon that has undergone a phase shift.

What this line of thought enables is the conception of a subject that is intermingled but still has a will of their own. Their volitional relation is built on a core loop meshed with a multiplicity of others who contribute with multidirectional influence and with varying degrees of intensity. In that sense, we are thinking *through* and *in* each other.

There is an overlap between several of the points I have elaborated above with regard to the sharing of consciousness. If we articulate Kojève's observation of the person as a microcosm, Lefebvre's idea of the joint efforts of many thinking individuals, Torey's endogram, Metzinger's phenomenal self-model, Baars global workspace and Hofstadter's intermingled strange loops then we can surely move towards something like a *social endogram* or *phenomenal social-self model*, or even perhaps the catch-all term *global correlate of social consciousness*. If our self-model contains, as it must, multiple fragments of other's selves and experiences as well as our own, then these will be manifest in our mental global workspaces. Such a model will grow in elaborateness and range, the greater the experiences and the extension of each consciousness. So, while the specific brain contains a specific consciousness with its neural correlates, at a collective level these are profoundly intermingled. This idea also speaks to Honneth's conception of care as well as the notions of becoming and collectivity framed in the previous chapter.

Emotion, as with so much else, is again of vital importance in realising the global correlate of social consciousness, in particular in translating this into collective action. Emotion is one key element that allows us to think and act together in real time. Emotional connections mean that when we think together we don't have to go through the detailed drawn-out process of thrashing out every decision or working through every other person's perspective. This is done in the laying down of what Damasio refers to as the 'somatic markers' that underpin the kind of self-reflexive decision making characteristic of deliberation and that allow for shortcuts in situations of intense speed and brevity of communication.

Damasio argues that somatic markers are emotional memories that are layered into our brains. As desires and preferences in response to external stimuli, emotions become embedded in our unconscious brain and are subsequently retriggered by particular situations or events. Somatic markers allow us to make decisions without trawling through every element and option, to act in ways that we may not always be fully aware of in our deliberate decision making. Given the speed of response that is needed for many decisions this is a basic necessity (Damasio 2010). The 'somatic marker hypothesis' is based on a combination of innate learning mechanisms, some of which became embedded in our unconscious, but are nonetheless learned and which are necessary for adaptation and survival in specific contexts. In some instances the actual decision-making process may be completely unconscious: '[t]he requisite knowledge was once conscious, when we first learned that falling objects may hurt us and that avoiding them or stopping them is better than being hit. But experience with such scenarios as we grew up made our brains solidly pair provoking stimulus with the most advantageous response' (Damasio 2006: 167). As a result, we are able to make myriad decisions on a daily basis without excessive cognitive stresses or huge dilemmas when weighing up all the evidence. Our somatic markers create emotional biases towards certain actions that are learned and incorporated into our very wiring. This has to do not only with purely unconscious actions, however, since it can also direct more traditionally 'rational' decisions. With any challenging, complex, or even seemingly straightforward decision, 'it will not be easy to hold in memory the many ledgers of losses and gains that you need to consult for your comparisons' (168).

The solution to this is for the more reflective conscious elements of the brain to categorise the benefits and risks of certain actions over time, and to link these into positive or negative emotional responses in the non-conscious regions of the brain, so that when confronted with familiar or even new situations we react quickly and seemingly instinctively, although in fact this can still be done rationally. Damasio argues that in a reasonably healthy society this is exactly how our brains should develop, so that '[s]omatic markers do not deliberate for us. They assist the deliberation by highlighting some options (either dangerous or favourable) and eliminating them rapidly from subsequent consideration' (174). Somatic markers 'are thus acquired by experience, under the control of an internal preference system under the influence of an external set of circumstances which include not only entities and events with which the organism must interact, but also social conventions and ethical rules' (179).

The shortcuts enabled by somatic markers reinforce the capacity for thinking and feeling together, the 'loopiness' of entwined consciousness and the social endogram. Acting through the brain's social endogram entails a social level that draws on shared somatic markers: the embedded knowledge, mores, habits and practices of others. Hence the situation in which long-term partners are able to think and feel through and with each other with a profound shorthand. This also applies at a wider social level with families, affinity groups and wider communities – helping to lock people together into intermeshed networks with shared perceptions and values. As Damasio argues, in neural terms somatic markers evolved in response to primary emotion, however 'most somatic markers we use for rational decision-making probably were created in our brains during the process of education and socialisation, by connecting specific classes of stimuli to specific classes of somatic state. In other words, they are based on the process of secondary emotions' (177).

To draw on Hofstadter's terms, our strange interlocking loops become ever more entwined though processes of repetition and mimesis, and they tie together interlocutors in loops of understanding and recognition. In Damasio's terms, there is a constant movement here between the three levels of the protoself, the core self and the autobiographical self. The fact is that we have wills, and can make decisions on the basis of reflective thought, but we are not subjects of complete 'free will'. Our will is 'steady and stable, like an inner gyroscope, and it is the stability and constancy of our non-free will that makes me me and you you' (Hofstadter 2007: 340). This flexibility within boundaries enables us to absorb and deal with affects, recognise where they are folded into conscience and consciousness, and act on them. As such will, volition, decision, as well as social intention, are meaningful aspects of the human situation.

EXTENDED AND SITUATED CONSCIOUSNESS

The intermeshing of strange loops, and hence the extent to which we 'live in each other', is always already integrated with and enabled by processes of externalisation, but in a technological society our social endogram is extended through our devices. The externalisation of our loops using technology to extend our thoughts, memories, desires and so forth will 'allow them to be shared by other people' (Hofstadter 2007: 231). Such extension allows for elements of loopiness to transcend time and space and to integrate with each other in ever more expanded terms; this is something that, while touched on by Hofstadter, is not developed much further. Indeed, in Hofstadter's work this technological aspect is an add-on, a supplement to our cognitive processes.

However, this extension of consciousness is of much greater significance when we reflect on the evolution of the human as, in many ways, always already a technological being – as we have seen with the discussion of Dasein and gadgets in Chapter 1. Humans are born in extremely physically vulnerable states, but this has the evolutionary advantage of allowing for high levels of brain development and plasticity. It is this vulnerability, and the need for social and environmental support, which in many ways continues throughout our lives, and ties us inextricably to growing interdependently with the world around us. We are not discrete autonomous beings, and this is certainly also the case with consciousness. There is a growing field of thought that sees consciousness not as something that takes place purely inside us and which is then extended and shared; rather, extension is seen as always already part of the whole composite of consciousness.

Alva Noë – a leading philosopher in the field of extended cognition – argues that consciousness does not actually happen in the brain, but rather in the dynamic process of the interaction between brain, body and environment. He tells us that 'to understand consciousness in humans and animals, we must look not inwards, into the recesses of our insides; rather, we need to look to the way in which each of us, as a whole animal, carries on the process of living in and with and in response to the world around us' (2009: 7). In that sense consciousness isn't something that happens to us, or emerges from us, but is something that we do actively as beings engaged in the world. The brain, rather than being the sole repository for consciousness – and as such being fully explainable with reference to the workings and configurations of sets of neurons, with an attendant describable 'neural correlate of consciousness'– is in fact more of a node and facilitator of consciousness; hence Noë proposes 'that the brain's job is that of facilitating a dynamic pattern of interaction among brain, body, and world' (47). One source of evidence for this is the increasing consensus around the plasticity of both child and adult brains. Noë makes the case that, contrary to some traditional views of brain development, we do not have specialist pre-destined neurons or zones of brain function that are specifically evolved to undertake certain tasks. For example, sight and navigational skills can emerge in atypical areas of the brain for some individuals with either brain damage or altered sensory inputs, indicating the great adaptability of the neuronal substrate of the brain. Plasticity suggests that '[s]ensory stimulation produces the very connectedness and function that in turn make normal consciousness possible' (49). In other words, it is stimulation by, and interaction with, the world that produces the experiences that bring consciousness to life, and indeed consequently the self. Noë also refers to the rubber-hand experiment described by Metzinger to support his position. That experiment manipulates an able-bodied subject, through the use of mirrors, into thinking that they can feel the touches made to a rubber hand; thus they experience a separate and inert hand as their own. Noë draws from this the idea that humans are 'dynamically distributed, boundary crossing, offloaded, and environmentally situated, by our very nature' (68).

The extension argument is an ontological one in that it is not merely that external elements enhance and extend consciousness beyond the brain, but that they are fundamentally interconnected with the development of consciousness in the first place, and the brain is only one aspect of the living situation in which consciousness and the self occur. Being human means that 'maturation is not so much a process of self-individuation and attachment as it is one of growing comfortably into one's environmental situation' (51).

We can posit that the objects, devices and tools that we integrate into our everyday life, and that to some extent become extensions of our physical capacities, are also very much integrated into us at a much more profound level. Our self-model, our subjectivity, is built around an assimilation of these objects on the social and cultural levels as well as on an individual level. Noë talks about the experience of living and working in a particular city where we are able to navigate with ease, where familiarity and routine mean that the spaces of the city themselves become part of our extended world, something we think with as we integrate the city's signs and symbols, its rhythms and patterns and language, into the boundaries of our own self, indicating again that '[w]e ourselves are distributed, dynamically spread-out, world-involving beings' (82). He argues that distribution is most striking when we travel or move into a new area or experience significant life changes; in such a situation we are thrown back on ourselves, being unable to rely on our taken-for-granted extended cognition. Acknowledging his Heideggerian influence, Noë suggests that 'The disruption that we feel when we travel is a giveaway of our usual unthinking reliance on that background of skills that make functioning in the world possible' (121).

Noë is not the only proponent of such views; his theory is a variation of a position that has been named *vehicle externalism*. Another advocate of this view is the philosopher Mark Rowlands, who in a similar vein to Noë has 'maintained that *some* cognitive processes do not merely depend on external processes but are *constituted* by them' (Elpidorou 2012: 146). Externalism is based on denying that 'mental or cognitive processes have to be identical to, or exclusively realised by, brain processes' (146). This claim is built on a view of the nature of mind which is careful to avoid essentialist or reductionist claims, including what Rowlands calls the view of 'the mind as a *bare substratum*' (2010: 10). There is no essence of mind that sits behind the collection of 'mental states and processes' that we experience as mind. The lack of essence echoes Hofstadter, who talks of consciousness bootstrapping itself through strange loops, with no need to rely on a mystical non-material or extra-material substance. Mind can perfectly well incorporate elements outside of the brain and body, and there is no hard and fast border between the internal and external aspects of mental states; again, this chimes with Hofstadter, in that mental states can, logically at least, be partially shared. Our capacity to adapt and absorb our environment into our brains and our self-schema is what gives us such a dynamic relationship to objects. The ubiquity of gadgets means they have surely become some of the most important modes of extension and situatedness for consciousness. The way we lock into, absorb or otherwise relate to gadgets as objects or things has the potential to frame the character of our current human condition.

THE PLASTIC BRAIN

The capacity of the human brain to absorb, adapt to or alter its external relations also becomes of great importance here. Brain plasticity is the relatively simple idea that human brains develop in ways that are not entirely genetically pre-programmed, that they respond and change in relation to input from their environment – not only in the sense of reacting or learning on the periphery but in significant ways in their material structure. This happens throughout childhood development, but also into and throughout adulthood, as the neuroscience specialist Moheb Costandi explains: 'The adult brain is not only capable of changing, but it does so continuously throughout life, in response to everything we do and every experience we have' (2016: 2).

Catherine Malabou transposes these scientific developments into a number of engaging cultural and philosophical claims. She starts by explaining that there are three aspects of plasticity: developmental, modulational and reparative (2008: 17). The first aspect relates to the brain's developmental flexibility and the extent to which, while not utterly shaped by outside forces, it does develop to a significant degree in response to its environment, which includes the forging of multiple new connections: 'a modelling of synapses or a mechanism of synaptic plasticity' (20).

The second form, modulational plasticity, is the most relevant for the discussion here. It describes the brain's capacity to interact with and adapt to its surroundings, and also to its own internal changes; that is, to respond to stimuli in more than a purely reactive way, with active self-modification, adaptation and agency. As Malabou explains: 'there is a sort of neuronal creativity that depends on nothing but the individual's experience, his life, and his interactions with the surroundings' (21), and she infers that '[p]lasticity thus adds the functions of artist and instructor in freedom and autonomy to its role as sculptor' (24). Plastic brains don't merely receive and absorb stimulation but 'have the power to form or to reform this very information', with the result that 'our brain is in part essentially what we do with it' (24).

The third form of plasticity, the reparative, refers to the brain's capacity for repair, specifically in the two categories of neuronal renewal and compensation for losses caused by lesions or other damage (24). The upshot for Malabou is that thanks to its plasticity the brain is an active agent in shaping the environment around it, and as such contributes to the collective shaping of human development.

If we combine this power of plasticity with shared extended consciousness and the social endogram, then we have a framework in which we can insert the gadget and start to consider more fully the nature of gadget consciousness. This framework is an ensemble that enriches and enlarges all the components into something more than the sum of its parts. It provides a picture that can resonate with the push for care, becoming and collectivity. These values can become ever more embedded in the gadget and the brain as they coevolve – gadgets through design and adaptation, and the brain through plasticity.

However, as with capital's tendency towards the gadget-object, so too does it wrestle with the plastic brain. Malabou draws compelling political conclusions from her exegesis of the facts of plasticity. She argues that the current formulation of capitalism prizes the capacities of flexibility and adaptability, and this means favouring certain modes of fluid and adaptive command. There is an attempt to leverage and direct the free agency of the brain into the service of capital by adapting and distorting the features of plasticity. As such, 'The biological and the social mirror in each other this new figure of command' (33). Capitalism has moved from a centralised hierarchical system of command to a decentralised one of localised tasks and temporary clusters of workers, cooperating to achieve specific ends, but interlinked within global networks. This understanding of plasticity echoes findings in neuroscience according to which the brain does not have any one command and control centre but in its plasticity is decentred. It operates on the basis of clusters of neurons firing together in varying and multiple configurations; this is an emergent rather than top-down process in which '[t]he primary qualities of assemblies of neurons are their mobility and then multifunctionality' (44).

The contemporary ideology of neoliberalism attempts to take this mirroring to justify its flexibility in light of plasticity and to naturalise its practices; hence both neoliberal, networked capitalism and contemporary neuroscience share a certain perspective that is '[o]pposed to the rigidity, the fixity, the anonymity of the control Centre' and supportive of 'the model of the suppleness that implies a certain margin of improvisation, of creation, of the aleatory'. As such, 'The representation of the centre collapses into the network' (35). Capital takes full advantage: 'The questioning of centrality, principal transition point between the neuronal and the political, is also the principle transition point between neuroscientific discourse and the discourse of management, between the functioning of the brain and the functioning of a company' (40). This crossover of the neural and managerial in the current economic order is an attempt to strengthen neoliberal hegemony by absorbing these neurological insights. Neoliberalism, however, does not offer real flexibility but an empty version that offers only malleability in a passive mode: 'Rather than displaying a real tension between maintenance and evolution, flexibility confounds them within a pure and simple logic of imitation and performance. It is not creative but reproductive and normative' (72). In this way, the logic of flexibility, adaptability, creativity and so on is recuperated and deracinated in the current order.

The gadget-brain ensemble becomes a key part of the challenge to a capitalist hegemony which insists that the only flexibility is that of labour and the only agency is that of individuals and markets. Malabou offers a powerful tool in this struggle by arguing for a distinction between malleability, flexibility and plasticity that allows for the reclaiming of the latter as a distinct and particular characteristic that need not, indeed should not, be associated with the flexibility of neoliberal capital. There is an analogy to be made between this reclaiming of plasticity and the reclaiming of the ensemble of the gadget-thing from the techno-capitalist dispositif. The need for a consciousness of plasticity is thus underlined, in the sense of both consciousness in and of itself but also of consciousness as associated with consciousness raising, in order to find ways of resisting co-optation and domination, separating flexibility from plasticity so as not to give 'too rigid prominence to flexibility, that is to say, to docility and obedience' (53).

Malabou offers a challenging dialectical view of the distinction of plasticity from flexibility, one that emphasises the conflictual and contradictory aspects of the dialectic. She highlights the contradictory tension between homeostasis, i.e. the brain's need to maintain equilibrium, and negation, i.e. the brain's need to negate itself in order to develop. As such, 'the individual ought to occupy the midpoint between the taking on of form and the annihilation of form' (70). This means that a struggle and a degree of resistance is necessary, in which there is no peace or resting point as such: 'All current identity maintains itself only at the cost of a struggle against its autodestruction: it is in this sense that identity is dialectical in nature' (71).

By integrating the ideas of the plastic brain, the social endogram and the gadget as thing, we can conceive of an ensemble that is radically receptive and adaptable to the outside world, but that also reshapes and reworks the world on an active conceptual and material level: conceptually through the levelling up of strange loops to the self, but then also materially by reshaping the world itself through the development of will and action. In doing so the gadget ensemble is locked into a relationship with the material world, not just in the sense of a tool but in extending and reflexively enhancing the brain and the self. In many ways brains are the ultimate gadget, reaching out to gather; the brain is the crux and crucible and nexus point, but existing in a constant condition of becoming as the body, the senses, the mind, the spatiotemporal flux of matter and energy, pivot on this point.

Yet is it meaningful to talk of the brain as gadget, or at least to make that analogy? Where does it get us? I would argue it takes us beyond the crude humanism of anthropocentric enlightenment but also splices a powerful agent and subject back into the insights of 'post-humanism', opening up a space for the human, which is inconceivable as isolated or primary. Is the brain then a 'thing'? In many ways we can argue that it is really *the* thing, the thing that activates and brings to life the things around it. While Damasio conceives of the core self as activated in the encounter with objects, we can go a step further and see our relationship to things as also springing from this encounter. We are activated as selves in our relation to objects, and as we incorporate them and entangle ourselves and our consciousnesses in and through them they take on a life of their own – they start to gather of their own accord as things, tying us to each other and opening our consciousnesses into each other. To be a little flippant, but also very serious, we can say that the brain is the *thinger*: the thing that makes things *thing*. But the thinger doesn't thing prior to, or in isolation from, the world of things with which it interacts and out of which it is constructed.

At this point in the argument we can also return to Hegel and the idea of absolute consciousness. As discussed previously, Hegelian absolute consciousness requires a totalising spirit; this is no longer a viable proposition however, and as Lefebvre argues, the question is actually about the interconnections between multiple minds. We have explored a framework for understanding precisely this interconnection – starting with the neural substrate of the brain, and leading to the evolution of the autobiographical self, with the idea of the strange loop of consciousness as a powerful bridge from the neural to the mental to the self and the intersubjective, and from there towards extended consciousness, and then back again to the remoulding of the self in the form of plasticity of the brain.

This trajectory leads us back to the question of class and political consciousness. As Malabou has intimated, capital is deeply implicated in how 'flexibility' and intersubjectivity are framed and shaped. This refocuses our attention back on the extent to which gadgets are subject to forces beyond the individual and the intersubjective – they are not neutral carriers of consciousness but material products of advanced capitalism that need to be actively engaged. When we conceive of the brain as a gadget, even a very special one that makes and alters itself, it must be understood that it is also a gadget brain, that it is formed and shaped and oriented by gadgets for good or ill, and as such the imperative becomes to understand more fully the whole dialectic of gadget consciousness.

4

Gadget Consciousness

In this chapter I am going to articulate the picture of the brain and consciousness with the concept of the gadget, in order to move towards a more developed description of 'gadget consciousness'. This will be a view of consciousness as a form of thinking together that is augmented and extended by gadgets, and that includes the sense of consciousness as both self-consciousness and collective consciousness. I will also address a more troubling sense of the term, namely that of a consciousness dominated by gadgets.

In order to get to that point, it will be helpful to look at what we can learn from previous theories of technology and augmented consciousness. The three most widely known approaches that I will touch on are: the singularity; collective intelligence; and multitude. All of these have useful insights to offer, but each is also distinctly problematic.

The first instance of a background theory that can help us think about gadget consciousness is that of the 'singularity'. Ray Kurzweil has posited this idea of the singularity, which is the moment of 'the culmination of the merger of our biological thinking and existence with our technology' (2005: 9). This merger has startling consequences in Kurzweil's thinking, wherein the human loses its meaning as we transcend into another realm of existence, 'overcoming the profound limitations of biological evolution' (21). This is underpinned by the assumption that '[m]any of the brain's characteristic methods of organisation can also be effectively simulated using conventional computing of sufficient power' (149). As such, this is a technological vision of artificial intelligence, which regards the computer as analogous with, and eventually a replacement for, the human brain. Kurzweil sees an exponential explosion of intelligence coming, that will only end when the universe 'has become saturated with intelligent processes and knowledge' (9). While this is a rather outlandish claim, it does expose a strand of thought that romanticises consciousnesses as a cognitive function alone – as somehow separable from the human body and social context – rather than something more fully embedded in the entirety of human existence.

The vision of the singularity dislocates itself entirely from the embedded political economy of everyday life, of the struggle for resources and the differences of sex, race, age and so on, all of which contribute to the constitution of identity and hence to our consciousness of self. In that regard there is here a return to a Hegelian drive of overcoming and a form of pure absolute consciousness that subsumes and replaces the discrete messy consciousnesses of the multiple. This perspective has an implicit unifying dream underpinning it, and, in line with this, suggests a somewhat totalitarian vision that would ultimately entail the end of politics. As well as being politically problematic, the singularity does not really offer much insight into the dynamics of collective thinking and acting as they have recently been seen across the world.

The singularity is also problematic because the place of technology is, and always has been, part of the make-up of the human. It is not a new additional element, the condition of which brings a release from our human bondage. This is not due to particular or insufficient forms of technology, but is rather an ontological constraint – we are always already 'post-human', and as such new technologies bring only a difference in scale and capacity, not fundamentally in kind. We see this overreach most starkly represented in science fiction, where scenarios can be played out to their most extreme. For example, in recent films such as *Transcendence* (2014), *Lucy* (2014), *Her* (2013) and *Ex_Machina* (2013), there is a representation of a singular bounded consciousnesses overcoming itself and merging into a grand super-consciousnesses. The human-computer hybridisation depicted in these films extrapolates from augmented and distributed intelligence into the monstrous realm, precisely as kind of 'becoming absolute' which is a variation of transcendent thinking – in this case it is a version of the singularity gone bad. However, these films are metaphors, extrapolating from existing technology and situations, to help us reflect on our current condition and think about futures that we wish to prevent.

Such depictions ignore the more modest 'joint effort of many thinking individuals' identified by Lefebvre (1968: 48). No doubt this is because such efforts do not make for such obviously dramatic storylines, and in some ways it is a similar story with theories of the singularity – which give a science-fiction spin to theories of technology. Such theories provide fantasies of limitless power and wealth to the disempowered and impoverished: Kurzweil promises we can have 'any physical product upon demand', that 'world hunger and poverty will be solved', and that 'we will be able to live as long as we choose' (2006). Here we are moving back into the realm of the religious and mystical, away from the fact that these are political and economic problems not technological ones. Gadget consciousness, by contrast, is focused on the other, 'joint effort', on the extended capacity to form collective will and solidarity in the world as it is, not the world as we might dream it to be.

This position also rules out the associated concept of 'strong artificial intelligence' (an AI that includes self-consciousness) as a model for or answer to gadget consciousness. A self-conscious AI is a distraction in that should it be achieved in any meaningful way, for example if the Turing Test were to be passed and even personhood conferred onto an AI, then it would simply reintroduce the same problem in two different ways. Firstly, the issue of the singularity would be reintroduced. Secondly, although it might empirically enrich our understanding of consciousness it would not necessarily address its ontological aspect. There may be a formal observable digital correlate of consciousness, and as such an understanding of certain states of consciousness and their material correlates, but the 'what it's like' to have that mental state would still only be available to the artificial consciousness, and describable and sharable in same manner as is currently the case. The insights offered by this knowledge are only conceivable to another observing consciousness; as such there is only ever a 'what it's like' to experience another consciousness' 'what it's like', and so the problem is deferred into an infinite regress.

The second conceptualisation of augmented consciousness, 'collective intelligence', has been developed most fully by Pierre Lévy. Lévy offers a self-confessed utopian vision, clearly influenced by the philosophy of Deleuze and Guattari. Lévy approaches collective intelligence as a molecular phenomenon that operates on a deterritorialised plane, decomposed from molar overarching forms of intelligence. His approach has the advantage of addressing directly the place of networked connectivity and intelligence, and the economic and social context and impact of this. He sees such molecular collective intelligence as generated through technological integration and the network form: 'it is a form of *universally distributed intelligence*, constantly enhanced, coordinated in real time, and resulting in the effective mobilization of skills' (Lévy 1997: 13).

Reiterating the immanent nature of this intelligence, Lévy tells us 'there is no transcendent store of knowledge and knowledge is simply the sum of what we know' (14). This means that, 'far from merging individual intelligence into some indistinguishable magma, collective intelligence is a process of growth, differentiation, and the mutual revival of singularities' (17). In that regard he offers a vision that is distinct from the totalising Hegelian view and is closer to the view of shared consciousness offered in the previous chapter. Lévy also incorporates a form of ethical demand into his notion of collective intelligence, which is slightly at odds with his Deleuzian ontology: he supports 'a new humanism that incorporates and enlarges the scope of self knowledge into a form of group knowledge and collective thought' (17). This is certainly a welcome perspective that reflects the power of digital communications to bring into relation huge numbers of thinking persons in much more intensive and fluid ways than was ever possible via mass media such as television or radio. As he points out: '[i]n immanent systems the mediator between the individual and the group is an electronic tool, held by thousands of hands' (68).

This is also paired with a certain political frame when Lévy argues that '[j]ust as nanotechnology can build molecules atom by atom, nanopolitics cultivates its communitarian hypercortex with the greatest attention to detail ... the members of the molecular community communicate laterally, reciprocally, outside categories and hierarchies, folding and re-folding, weaving and re-weaving' (55). He also offers some reflections on how collective intelligence changes the character of labour. This is familiar territory, and he makes the case for the ever-increasing significance of the value of knowledge and the creative and communicative capacity of workers, in particular intensely cooperative groups who are able to move and think quickly. He argues that:

Given equal material resources and similar economic constraints, victory will be claimed by those groups whose members work for their enjoyment, learn quickly, live up to their commitments, respect themselves and others, and move freely throughout the territory rather than trying to control it. Those who are the most just, the most capable of fashioning a collective intelligence together will succeed. (32)

While similar statements might be found in recent business literature, there is a properly progressive, almost utopian, aspect to this thinking.

Lévy goes on to advocate not a simple road to greater profit-making, but the possibility for socialising work: the 'day when the new proletariat attain self-awareness, it will dissolve itself as a class, it will bring about a general socialization of education, training, and the production of human qualities' (37).

There are useful insights to take from this, and to take forward in thinking about gadget consciousness; however, Lévy's approach reintroduces both transcendence and capital by the back door. In the first place, by overemphasising the place of emergence, his view of collective intelligence stresses precisely the *collective*. That is, the collective becomes the agent itself rather than retaining the sense of being in dialectical tension with the individual, or dividuals, to use the language of Deleuze. Ironically it is with the insistence on staying at the level of the molecular – although recognition of this level is vital – that the transcendent is smuggled back in. As Lévy tells us, while 'the thought of individuals is discontinuous because they sleep, grow ill, tired, or take vacations, the collective intellect is always alert'; as if this weren't enough, he goes on: by 'combining thousands of intermittent flickering rays, we obtain a collective light that shines continuously' (107).

There are also problems with the view of labour presented by Lévy. Rather than offering a break from, or even an overcoming of, labour, collective intelligence repeats important aspects of market logic, for example the idea that what collective intelligence represents is an 'expanded notion of liberalism' (34). This view rests on a curious hybrid position, wherein there are elements of cooperation but they are still framed by the logic of exchange; as such, '[e]ach of us would be an individual producer (and consumer) of human qualities in a wide variety of markets or contexts' (34). Lévy doesn't see this is capitalistic, believing that 'no one would ever be able to appropriate the means of production exclusively for their own use' (34). Yet this is still completely within the aggregative logic of market 'choices' that constitute the 'invisible hand' or the marketing vision of the 'wisdom of crowds'. Lévy must therefore see this as being in line with the idea of the proletarian class 'dissolving' in a moment of self-awareness. However, such attempts to overcome ideas of class ignore antagonism, or any idea that class composition can be tied to distinct contradictions of interests. As such Lévy's position can be read as having an idealist ontology, in line with his utopian self-proclamation. His view of a new form of democracy, in a society dissolved of classes, reflects such an idealist perspective; and even the neoliberal language of games is employed: '[t]his new democracy could take the form of a large-scale collective game in which the most cooperative, the most urbane, the best producers of consonant variety would win' (68).

The approach of collective intelligence has a lot to offer in the attempt to solve the problem of thinking 'together'. It recognises the multiple character of the collective, offering a compelling vision in which '[m]olecular politics resist the temptations to organize through separate entities. It plunges molar forms of organization into the cycle of collective intelligence' (73). But, without an account of the material underpinnings of consciousness and of the systems of communication that establish commonalities, it becomes too abstract and fails to engage with the material underpinnings of everyday life. Collective intelligence misses the necessity that the collective include both us and something beyond us, something more than the sum total of our molecular actions, which feeds back into longer-term allegiances and political commitments. Without larger group allegiances, interests and intentions there is no politics. Without politics there is a collapse towards the totalising and market-oriented underpinnings of the current fetishisms of neo-liberalism: infinite flexibility, plasticity and fluidity. Market-framed deterritorialisation ends up operating in the interests, if not the name, of capital.

The third recent attempt to grasp the dynamics of the collective is the development of the concept of *multitude*, which in many ways addresses a number of the concerns discussed above, but is much firmer in its political objectives and recognition of the tension between singularity and collectivity. Multitude is a formulation to understand the technologically augmented collective that is still in line with an idea of 'distributed consciousness', and supports an immanent ontology of the multiple, but within a firmly materialist and anti-capitalist mode.

The idea of multitude was originally developed by Baruch Spinoza but then taken forward in the tradition of post-Fordist, autonomist thinking. The autonomist analysis of capital accounts for digital and machine augmented collectivity, while maintaining awareness of the individual within the collective as a 'singularity' (though not in Kurzweil's sense). In so doing the autonomists neither fetishise individuals nor meld them into a mass.

Multitude is that which emerges when intensive modern communications enable singularities to come together and act in concert towards common ends, contributing their ideas and labour to a collective practice.

It is the digitisation of the economy and the development of global data flows that have provided the space and capacity for the multitude to act. In particular it is the distributed nature of these networks that allow the multitude to use technology to cooperate, turning towards each other in dense interaction, and away from capital imperatives. Thus the multitude recalibrates the technology towards its own ends. This means the development of a new kind of entity that is able to act collectively and creatively in an almost spontaneous manner. In doing so the multitude creates a new political subject.

This is not a technological determinist view – the seeds of the multitude have always been there, but are enhanced by networked devices. Multitude, says Paolo Virno, is 'a fundamental biological configuration' and what lies at its core is 'the publicness of the intellect' (2004: 94). In this regard, the concept offers an explanation of the capacity for common intelligence in action. It also allows for an analogy between the multitude as a social body and the biology of the brain. The jump from the neural networks that comprise the material substrate of brains to the social organisation of group intelligence is clear.

This line of thought is followed by two of the best-known thinkers to emerge from the autonomist tradition, Michael Hardt and Antonio Negri. In their book *Multitude* they argue that research in neurobiology shows that 'mind and body are attributes of the same substance and that they interact equally and constantly in the production of reason, imagination, desire, emotions, feelings and affects' (2004: 337). From this observation they infer that '[t]here is no one that makes a decision in the brain, but rather a swarm, a multitude that acts in concert … the human body is itself a multitude organized on the plane of immanence' (336).

The analogy is then made between the brain and the social body as distributed systems: 'The brain does not decide through the dictation of some centre of command. Its decision is the common disposition or configuration of the entire neural network in communication with the body as a whole and its environment. A single decision is produced by a multitude in the brain and body' (338). This logic is then easily scaled up to a global level with the huge expanse of digital communication networks and the attendant expanse and intensification of connections between brains. The key characteristic of these networks is their distributed topology; that is, they are many to many, all points connecting to all other points via numerous possible routes. Thus there is commonality with the concept of multitude, comprised of 'The innumerable and indeterminate relationships of distributed networks' (113) that produce the '[d]ynamic of singularity and commonality' (198) and a 'social subject and a logic of social organization' (219). This social subject is intimately entwined with distributed communication, and the notion of 'emergence' is implicitly used to explain the process by which collectives make decisions and act. Emergence is the mechanism whereby the combination of numerous small decisions and processes aggregate to produce effects that can be considered to express a kind of group intelligence.

This idea of dynamism is an attractive one with which to supplement the idea of the intention and the social endogram. However, there is a significant problem as far as emergence and decision making in the multitude goes – to put it bluntly, people aren't neurons in a brain. The brain produces thought as the result of billions of neurons generating trillions of connections, emerging out of the operations of multiple interconnected brain regions firing in synchronous waves of electrical and chemical actions. The connections are complex enough to create a first-order awareness which is then able, via the incredible density of the cerebral cortex, to loop back on itself to capture a picture of its own operation – the endogram. The brain is then able to create a set of priorities in the second-order 'global workspace', which manifests as will and develops into intention and action. Even if all human brains were directly linked via electrical connections, there are not enough humans on the planet to generate such a global workspace effect in the multitude.

There can then be no such equivalent correlate consciousness in the multitude. The closest processes are language-based agreements, norms and social rules that are established over time, but this is not the kind of spontaneous live decision making and agency that is ascribed to the multitude. The 'democracy' of the multitude, in so far as it tries to reproduce the logic of the brain, is an aggregative one that does not involve loopiness at a fast enough and intense enough rate – the multitude cannot be said to think, certainly not with regard to any kind of second-order intentionality (that is, with an awareness of the processes of decision making and the consequent conscious choices therein). Decisions may well get made at the macro level of the multitude as a process of aggregation of smaller individual decisions, but the feedback process can never take place at the rate required to generate reflexive decisions from the totality of the multitude. The objection could be made here that many decisions in the brain are unconscious, viral and affective in just this way – and this is a fair and useful insight. But the brain does have the added capacity for the self-conscious grasping of the impact of those unconscious tendencies though the global workspace. Self-conscious grasping is not a transparency that eliminates the unconscious as such, but a volitional capacity to insert conscious direction into action. The capacity of the brain to stop actions that are underway midstream or to change their direction, in more than just a reactive way, goes beyond that of the multitude.

If we consider the gadget as the point of interface in the multitude, then the aggregate of gadgets and brains does not and cannot make decisions with the degree of capacity or flexibility or volition that the brain itself does. It cannot be said to be conscious in any of the usual senses of the term, and in that regard we cannot see gadget consciousness as simply an equivalent of the multitude. Gadget consciousness contains the added capacity to create social intention. We can hypothesise that brains that are part of gadget ensembles are conscious in ways that are enhanced compared to non-gadget brains. Gadgets enable a harmonisation of the polyphony of voices, and as such there is a resonance between them; this can then be captured within the individual 'loops' of consciousness that Hofstadter describes. So it is that each interconnected brain carries a fragment of all the others. From this base, the emergence of group decisions can happen and the more intensive, the more embedded, those shared loops are the more the level of gadget consciousness can be raised – in that sense we can talk realistically of collective intention. This is the levelling-up moment of the strange loop enfolded through gadgets. This intensified and extended phenomenal social-self model, or gadget-augmented social endogram, creates a mode of extended thinking and empathy. In that regard it is something humbler than transcendent pure thought, but still active in real time. It creates an overlapping multi-intersubjective loop that would not be possible without the extended forms of communication offered by gadgets.

In developing the definition of gadget consciousness, we can take from the previous chapter the notion of the social endogram being extended and situated in and through technical devices, and we can frame the context of that extension though the notion of the gadget-thing. In that process of augmentation and extension we can take from Kurtzweil the broad sense that there is a merger of biological and technological capacities. But we should reject the possibility that this merger will entail or become a 'singularity' or that this is in any sense desirable. We can take from Lévy the idea that the pooling of capacity via digital communications can intensify cooperation and inaugurate forms of group intelligence, and that the multiple and molecular character of this cooperation is likely to change patterns of labour and socialisation. But we should reject the transcendent tendencies in this thinking that cause it to drift towards a Hegelian notion of the absolute mind. We can take from Hardt and Negri the idea of the intensive cooperation of the multitude and the notion of the multitude's resonance leading towards forms of dynamic emergence with political significance. But we should stop short of the idea that this is somehow equivalent to a conscious reflexive decision, and therefore that that emergence trumps cosalience.

While we have been focusing largely on what might be called the real-time aspects of gadget consciousness, there is a need to address another key feature that must be part of this account, and that is memory – which again entails a great deal of ambiguity for gadget consciousness.

GADGETS AS MNEMOTECHNOLOGY

The concept of gadget consciousness can be deepened by way of Bernard Stiegler's theory of grammatisation, which he defines as 'the history of the exteriorization of memory in all its forms: nervous and cerebral memory, corporeal and muscular memory, biogenetic memory' (2010: 33). Grammatisation in Stiegler's account tends to the solidifying extension of expression and ideas into material forms of greater duration. John Tinnell describes the overarching process of grammatisation as one wherein 'a continuous flux (e.g., speech, the body, the genome) becomes broken down into a system of discrete elements' (2015: 136). This includes writing, which generates a form of artificial memory, what Stiegler calls tertiary retention – a device which stores such retention becomes a mnemotechnology. By definition, everything digital fits into this description of grammatisation. Almost all digital communications are also stored, somewhere or other, and thus also contribute to tertiary retention. As a result, whenever we communicate via gadgets we also lay down a digital memory and extend the life span of that moment into the technology and beyond.

Stiegler's account of grammatisation is helpful in thinking further about the ambiguity of gadgets. The material forms shaped by grammatisation are referred to by Stiegler as the pharmakon. Pharmakon is a term taken from Greek and means something which is both poison and cure.

Stiegler employs Donald Winnicott's concept of the transitional object as the first pharmakon on the path to adulthood – the object (blanket, teddy, etc.) that allows the transition from total dependency on a parent to separation and autonomy. The object is what allows for the transition to adulthood, but within it lies the danger of excessive attachment, dependency and the destruction 'of autonomy and trust' (Stiegler 2013: 3). This is the double-sided nature of the pharmakon: we get very attached to and invested in such things, which absorb our desires and support our sense of security and worth, but which also undermine and cause the self-same need.

Writing as artificial memory is also such a pharmakon, in as much as it operates as a salve; it allows cultural memory to be extended and shared, but also, according to Plato, it decays autonomy of thought. Taking his lead from Derrida, Stiegler tells us that 'while Plato opposes autonomy and heteronomy, they in fact constantly compose' (2). The digital pharmakon, according to Stiegler, is the extension of this logic to the entire field of the human body, including in cognitive capitalism wherein 'those economic actors who are without knowledge [are so] because they are without memory' (2010: 35). Memory loss is the essence of contemporary proletarianisation, extended into the realm of consumption in which our 'savoir vivre', knowledge of how to live, is forgotten. We are no longer able to remember how to think because our lexicon is proscribed by the absorption and pre-emption of social networks.

In many ways we can see our relation to gadgets as objects as a clear example of such a proletarianisation process, as 'hypomnesis' (derived from hypnosis, referring to an empty circulation). We are thus reminded of Jodi Dean's description of the flow of communicative capitalism as simply 'drive', in which messages circulate without ever getting where they are meant to go.

Yet against this proletarianisation perhaps there is a gain, even in Stiegler's own thought. The therapeutic or individuating elements of this process can be found within the extension of gadgets from an immediately bounded form to one that connects brains together. As soon as gadgets go online to become mobile communication devices their capacities are exponentially increased according to the number of connected gadgets. Thus, there is a new relation to mnemotechnology which entails an increased velocity, multiplicity and ubiquity of grammatisation and tertiary memory. Attending to a gadget places the user in an immediate relationship to the aggregate of the 'just passed' and the 'passing through', but which also extends backwards and outwards beyond the individual brain and act of expression. With gadget use there is a proximate moment of shared expression, but also one that is placed in a cultural memory. As such, our connection to gadgets is a mnemotechnology par excellence, in that it augments human retention in the way the data is inscribed in prosthetic storage. In certain contexts, such as an open access social media app, this memory can then become accessible to any other user of the network, becoming absorbed into their device and therefore also potentially available to their social endogram.

While the long-term availability of tertiary memory extends the social endogram in principle, the vast majority of these memories are only fleetingly attended to, so in many ways it replicates the temporal form of speech. While pre-gadget, pre-networked media forms also enfold their own recording and temporal extension – print media in libraries; film and television in video archives; sound and music in libraries – digital grammatisation is much faster, much more easily overwritten, altered or responded to, and as such more fragile.

Thus, again we see the contradictory tendencies within our gadgets. In the first instance they are made out of a system that is deeply rooted in capital, they capture our memories and insinuate themselves into our most intimate relationships. Yet they also bring us together in ways that can augment and enhance our thinking, knowledge, empathy and collective volition, precisely by tapping into this common pool of experience, understanding and desire.

The concept of the somatic marker can play an important role in enriching this more active and positive understanding of our relationship with gadgets. As was discussed in the previous chapter, much of what is processed by the brain is done so unconsciously. Many choices and actions are therefore based on responses that occur before we are aware of them, often directed by somatic markers that have been embedded over longer periods of time. Stiegler argues that we pre-empt our volitions with the already inscribed emotional or affective layer, protending beyond the immanent into the virtual. Protention, while beneath the surface, is still – according to Stiegler and others – a fundamental element of consciousness. It pushes forward into the brain's awareness of continuity, contributing to its affective reactions, and is central to projection and risk. This aspect of protention therefore is a fundamental contributing element of volition as it rises into consciousness. Volition is the active conscious aspect of willing, and as such requires an act of protention to underpin it.

Protention, as Stiegler describes it, will also be inscribed in the flow of gadget consciousness, but written into the cognitive process that precedes and frames it. This allows us to think of the gadget-brain 'assemblage' or 'dispositif' as not just a real-time process of becoming, a tool or simply a medium in the linear sense, but again as something that gathers, orders and processes not just the present but the past and the future. When Maurizio Lazzarato reflects on dispositifs he sees them as 'machines for crystallizing or modulating time ... capable of intervening in the event, in the cooperation between brains, through the modulation of the forces engaged therein ... Consequently the process comes to resemble a har-monization of waves, a polyphony' (2006: 186).

This is an excellent framework within which to consolidate the place of the gadget as part of just such a polyphonic dispositif that would include the 'gadget brain'. In the first instance the place of the gadget in 'crystallizing or modulating' time is reflected in its grammatisation of the immediate into a circuit that reframes the present moment in a series of ripples and echoes, which resonates in the protentions of the interlocutors. This organising of thoughts and affections in a temporal multiplicity crosscuts events, to the extent that the event is conceived as something new that enters the world. So it is that the permanent process of sharing, narrating and modulating changes the shape of events from pinpointed moments of impact into flat plains, or membranes, that intersect with mental events. The brain-share, or what might be called a 'brane' of brains, unfolds both spatially and temporally, but within the limits already defined.

This 'brane' of brains can be understood in Lazzarato's terms precisely as a 'harmonization of waves, a polyphony'. The dispositif produces this modulated consciousness as part of a distributed condition that provides for a cooperation between brains via a multifarious looping. This looping allows for collective protentions, which as part of a process of becoming shape action and can be understood retroactively as volition. This, again, points us towards resonance as a powerful force in gadget conscious-ness, and to the idea that gadgets, as forms of mnemotechnology, can contribute to the thinging of gadgets.

It is therefore clear that this technological change needs to be understood together with notions such as 'noopolitics' and 'neuropoli-tics'. Lazzarato captures this very well when he tells us that 'noopolitics commands and reorganizes the other power relations because it operates at the most deterritorialized level (the virtuality of the action between brains)' (187). However, again, the danger is not far away and is well-defined by Stiegler:

> When technologically exteriorized, memory can become the object of sociopolitical and biopolitical controls through the economic investments of social organizations, which thereby rearrange psychic organizations through the intermediary of mnenotechnical organs, among which must be counted machine-tools. (2010: 33)

Here too we find the potential for proletarianisation, in which gestures, knowledge, skills become – in the medium and long term – separated from the bodies and brains of workers and turned into mechanisms that make them forget. There is therefore a real possibility that the short-term resonance and collective volition will enter a distorted and heightened state, with a rather unpalatable after-effect, in which the memories remain only as commodified digital data. The question is therefore whether gadgets will remember, and think, for us – and in so doing relieve us of that know-how. There is a scenario wherein gadget consciousness is reduced to a state of always already forgetting:

> The proletarian, we read in Gilbert Simondon, is a *disindividuated* worker, a labourer whose knowledge has passed into the machine in such a way that it is no longer the worker who is individuated through bearing tools and putting them into practice. Rather, the labourer serves the machine-tool, and it is the latter that has become the technical individual. (Stiegler 2010: 37)

Again, the pharmacological character is apparent. According to Stiegler, 'the Internet is a pharmakon', blurring both 'distributed' and 'deep' attention (Crogan 2010: 166). It is a marketing tool par excellence, and here its capacity to generate protention operates to create not only a collective 'volition' but a more coercive collective disposition or tendency, that is the unconscious willing or affective reflex. This is something more akin to what Richard Grusin refers to as 'premediation'. In premediation the future has already happened, not in the actual sense, but such is the preclusion of paths of possibility that we cannot conceive otherwise. Proletarianisation operates in this way through platforms that turn gadgets into data-gathering objects. Here communication is not a thoughtful exchange between skilled interlocutors but a habitual response to a standard set of pre-digested codes (in the sense of both programming and natural language) ready to hand to be slotted into place. In this scenario, somatic markers are also turned to a problematic use when they become saturated by ideology as they sink into the brain's affective reflexes. In that regard, there is a proletarianisation of the prosumer – in a form we can call gadget false-consciousness.

It is here that we can return to the significance of brain plasticity as something of a battleground. As well as being well designed for the production of strange loops and ego tunnels, the brain is also evolutionarily predisposed to shift attention towards new information, and as such there is a tendency for its sustained wide internal connectivity to break down quickly – in short, concentration is easily broken. Neuroscientist Adam Gazzaley and psychologist Larry D. Rosen describe such interference as 'a fundamental vulnerability of the brain' (2016: 3). When setting about achieving a goal, the brain is very easily distracted. As Gazzaley and Rosen state, the brain is 'undeniably the most complex system in the known universe' (8), but all such systems have their weaknesses. When we add to this the context in which the brain has to function – modern complex cultures where goals are always crosscut with a multiplicity of reasons, relationships with others, systems and multiple layers of meaning – it becomes an immense challenge to keep track and remain focused. We are always contending with the limitations of our brains.

Brains are bad at 'cognitive control abilities', which means we have a limited ability to 'distribute, divide, and sustain attention; actively hold detailed information in mind; and concurrently manage or even rapidly switch between competing goals' (9). Since we can't keep much information in mind at any one time or with any sustained duration, there is 'a tension between what we want to do and what we can do' (10). Beyond this limitation we also have an evolved proclivity towards distraction: humans 'engage in interference-inducing behaviours because, from an evolutionary perspective, we are merely acting in an optimal manner to satisfy our innate drive to seek information' (13).

This is apparent enough in any time period with relatively modest sources of information or novelty, but in a technologically saturated and information-rich era with ubiquitous and constantly connected devices, the potential for interference and distraction is huge. So it is that 'many technological innovations have enhanced our lives in countless ways, but they also threaten to overwhelm our brain's goal-directed functioning with interference' (4). They also have the power to inflate these behaviours by ramping up 'their influence on internal factors such as boredom and anxiety' (13).

Into this comes not only the technology itself, but the parameters and needs of capital. In this context, distraction and attention are two aspects of the same process: the grabbing of attention for long enough in order to create an awareness of products and stimulate desire, and the distraction from deep reflection and contemplation in order to limit critical engagement, broader dialectical thinking or the forming of deep and enduring relationships.

It is therefore capital's purpose at the neurological level to install somatic markers that associate good feelings with products, services and needs that suit the extraction and generation of surplus value. The brain is shaped on an ongoing basis by the constant bombardment with stimulation that trains it into craving only short-term immediate gratification, undermining its more developed reflective and cognitive controls. The aim is therefore to short circuit the brain, to break down the strange loops into disconnected bursts of desire and gratification – to transform the ego tunnel into something far less capable of consciously 'loopy' decisions and shared or 'resonant' loops. This is the battle for attention, as the primary commodity of the digital age, and the route to priming the consumer for this is directed via the control of gadgets towards moulding the plastic brain.

However, it needs to be reiterated that this is not the final say on the matter; that where there is volition, and in particular collective volition, there is also the possibility of a gadget consciousness. This will involve leveraging self-reflective consciousness – the strange loops – towards shaping the plastic brain to respond differently and lay down somatic markers of trust, commonality and solidarity that shift our affective responses and social ties in a different direction, away from the imperatives of capital. The need is therefore to design, and comport ourselves towards, gadgets as focal things – not as objects that fragment, distract, interfere and gratify with short-term rewards.

The claim here is that by taking hold of this gadget consciousness, and transforming it into an active collective volition, we stand the best chance of finding 'a political will capable of progressively *moving away from the economico-political complex of consumption* so as to *enter into the complex of a new type of investment* … or, in other words an investment in common desire' (Stiegler 2010: 6). In its most simplistic form this requires a new political economy of commoning, wherein gadgets contribute to a broader augmented volition that is not captured within communicative capitalism, coded to turn volition into capital, but rather turned towards a gadget consciousness as a common desire for a different path. What is being described here is effectively the need for class consciousness.

TOWARDS GADGET CLASS CONSCIOUSNESS

We can imagine human brains as one element of a wider media-technology hybrid wherein human intelligence, consciousness and memory are augmented and extended through gadgets. We've already dismissed, at least partially, three of the usual positions on this topic, that of the singularity, collective intelligence and the multitude. I have argued for the efficacy of gadget consciousness, and the tensions between the kinds of consciousness that are suggested by an orientation towards gadgets as things or as objects. To further the case against the object orientation, it is worth exploring an even older conceptualisation developed by Karl Marx, that of the *general intellect*. Although it was influential in the formation of the concept of multitude it is well worth returning to this idea in and of itself, not least because this will allow us to reintroduce an economic aspect to recover the gadget brain as a productive force.

Marx's notion of general intellect captures dynamic elements of knowledge and know-how (1973: 706). Although writing way before the advent of modern computers, Marx was well aware of the development of automatic machinery, the work of Charles Babbage and the drive towards factory automation. He sees machines as '*organs of the human brain, created by the human hand*; the power of knowledge, objectified' (706). These extracts are from Marx's famous 'Fragment on Machines' (690–712), in which he explores the development of fixed capital – that is, the means of production itself – as a necessary element in the development of capitalism more broadly. He argues that this is possible only at the moment when surplus labour is developed to such a point that it can be expended in non-vital labour; that is, when the efficiency of production reaches a point at which necessary labour time (that required for the mere reproduction of capital) is significantly less than the actual amount of labour power that needs to be purchased by capital to maintain equilibrium. The creation of surplus value (profit) from this surplus labour can then be reinvested as fixed capital, for example in improved machinery, to further boost efficiency and so create another cycle of profit and investment. The development of machinery, at a certain point in the advancement of capital, thus increasingly relies on the absorption of more of the technical skills and knowledge of living labour, petrified into dead labour (fixed capital).

As Marx argues in *Capital*, 'Capital is dead labour, which, vampire-like, lives only by sucking living labour, and lives the more, the more labour it sucks' (1976: 342). In the 'Fragment' Marx had already theorised the role of automaticity, arguing that, unlike a tool, 'which the worker animates and makes into his organ with his skill and strength', the machine captures the worker as its own appendage, with capital therefore 'acting upon him through the machine as an alien power', turning him into 'a conscious organ ... subsumed under the total process of the machinery itself' (1973: 693). This highlights for Marx the vital importance of knowledge, and social knowledge in particular, in as much as capital depends on '[t]he accumulation of knowledge and skill, of the general productive sources of the social brain' (694). The significance of social knowledge is profound, in two distinct directions: firstly, in the deepening of alienation and the capture of living labour as an appendage of capital; but secondly in the possibility of the collapse of capital itself and the potential to use machines to free humans from toil. These possibilities are highlighted in the dynamic of ever increasing amounts of investment in fixed capital (machinery), achieving a point at which – given the elimination of labour, or its radical reduction – value can no longer be extracted from the process by capital. At this point social reproduction could be maintained with minimal labour, allowing the possibility of the full development of human beings. Machinery, Marx points out, still retains its use value even when it ceases to function as fixed capital, and as such it does not follow that 'subsumption under the social relation of capital is the most appropriate and ultimate social relation of production for the application of machinery' (700).

While there has been a great deal of debate over the central point of controversy in this approach – namely, whether machines, in particular automated computers, can create value (Caffentzis 2013) – this is not the pivotal point for my argument. Rather, it is the contradictory character of machinery that is of interest here: on the one hand the subsumption of collective memory and knowledge, or general intellect, to fixed capital, and on the other the capacity for liberation that is simultaneously inherent in the mnemotechnology, in so far as the latter contains the seeds of its own undoing when memory becomes common and breaks free from the fetters of 'big data'. This doesn't produce the collapse that Marx had considered in the Fragment, but it does suggest the power of an active general intellect to recuperate its own memories and skills.

The novelty that I am proposing here is that Marx doesn't really consider the use of machines on an individual or small-group level. In our current techno-capitalist configuration the efficiency with which mechanisation has evolved into miniaturisation, cost-reduction and mass production puts a computer on every desk, at home as well as in the office, and even into every backpack and pocket. The technology – certainly where cognitive labour is concerned – is thus now embedded in social life and the coordination of all activities. As such, harnessing the technology outside capital, beyond the assigned role of 'mere' worker, is the vital element. Marx's view applies here more than ever, as manifested in the evolution of commercial social media platforms. Platforms aimed at management and control constitute the struggle by capital to block the transition to the *in-dependent* worker that the shift towards personal technology threatens.

Gadget technology has become increasingly sophisticated – capturing movement, consumption of media, fingerprints, facial recognition and so on. However, even when capital attempts to valorise this captured data it cannot fully enclose it – it leaves open the possibility of memories as common means of becoming collective. The excess of social media allows something like an *active* general intellect to emerge that is the counter-force of reification. Inevitably this means users negotiate a contradiction – the challenge of not submitting passively to capture by the power of recursion, but rather nurturing its shift to reflexivity. Gadgets are then a site of contest between the enhancement of memory as extended brains and cooperation, and capital's need to increasingly fix and control, to direct and commodify.

This then adds to the sense in which the gadget has at its core a tension that is its 'pharmacological' character. The curative aspect of the pharmakon in this instance is the possibility of an active general intellect, and that is close enough to be incorporated into what I've otherwise called gadget consciousness. With the elements of archive and memory in particular, this consciousness can also be conceived in the sense of class consciousness – something like a shared concept, idea or identification amongst a specifically bounded group in opposition to a prevailing power structure. In a political context we often hear of 'consciousness raising' as an aim, but we can conceive a more profound and dynamic concept of class consciousness unfolding that entails a shared sense of self, value and place manifest in micro-organising, coordination and collectivity, and which results in forms of concrete praxis.

This consciousness suggests a possible evolution in the struggle for new class relations. Such a struggle can entail emotional, affective and linguistic exchange and overlapping interests in what might be called movements, counter-publics or issue networks – or, to use my own term, 'quasi-autonomous recognition networks' (Hands 2011). These clusters can evolve within the frame of production of social life towards a capacity for class action, underpinned by the confidence in a solidarity to come. We have seen this in numerous examples: in the actions of UK Uncut and other such groups and movements around the world, most significantly in the multi-media augmented protests that clustered in Tahrir Square, Zuccotti Park and beyond. Such movements are profoundly embodied processes, happening in real places and to real bodies, involving fingers on the keyboards of devices, the buzzing of phones as messages are registered and rebroadcast, and also vitally the layering of experiences of sharing spaces as collectives. Whether the new capacities afforded by gadget consciousness will flourish, or whether constituted power will succeed in squashing them, remains one of the key political questions of our time.

DOES TWITTER THINK?

I have focused so far on the rather abstract notion of the gadget, but it would be helpful now to make this more concrete in order to illustrate and develop the preceding claims. One specific instance of the concrete presence of gadgets in everyday life and in the shaping of our brains is social media. The proliferation of social media, and especially their rapid shift onto diverse platforms, in particular to 'apps', means that they have become generally highly mobile, always connected, and operate through very sophisticated interfaces designed for maximum ease of use.

For a significant number of users, social media platforms have become a constant accompaniment to everyday life, the locus of a permanently unfolding self-narrative.

Since its inception in 2007 one of most popular and influential of these networks has been the micro-blogging service Twitter. Twitter is a gadget, in the sense that it is part of an assemblage that when in operation turns the whole device, whether a smartphone, tablet or computer, into a micro-blogging machine. As I argued in Chapter 1, the flexibility of a digital gadget is precisely that it can become whatever programme is running on it: just as a smartphone becomes a spirit level, so it can become a micro-blogging machine, and so too Twitter becomes a gadget. If we are also to define the Twitter-gadget as something that orients itself towards (at least the possibility of) thinghood then we can ask: how might Twitter thing? The Twitter interface is an opening onto the stream of thought in the forms of bursts of expressed consciousness, compressed into 280-characters. These thoughts are, of course, expressed in the form of language, but such snippets lend themselves to a rather different process of consumption than other long-form locutions. The ease of the interface, and the specificity of the action – of tweeting or scrolling through a feed – is the key here. The limited length of the tweet ensures that nothing takes more than a brief moment of attention and that it is possible, and routine, to process many messages and to communicate with multiple interlocutors, if not simultaneously then far faster than was possible in previous applications or technologies.

It is the speed of Twitter that creates an imperative to respond quickly and to assimilate vast amounts of information, to sort the agreeable from the disagreeable, divide that which should be ignored from that which should be responded to, and indeed that which calls to be acted upon. This produces a form of distributed attention, creating a wide zone of social awareness, in which the brains of Twitter users are able to process and respond to the perspectives of others almost instantly. Twitter's brevity and speed are therefore its most important affordances.

The speed of the feed means that it soon becomes impossible to see the flow as anything but multiple fragments. This fragmentary character is also intensified by the inevitable limitation of the number of accounts being followed by any one user. These two elements, which we can think of as axes on horizontal and vertical planes, are those through which selection and time cut. These cuts create the specific experience of each user, but then create multiple overlaps between users. We can also see these as integrated into further affordances of the gadget, as mobile, intimate and always on.

Such a platform design produces an ever-greater distribution of attention and temporal and spatial fragmentation – a combination of centrifugal and centripetal forces. Yet it is possible to alter one's orientation to the stream by focusing only on a limited number of feeds, or to use the organising tool of the hashtag. By identifying and clustering tweets with specific markers they become visible as a stream and the focus of all the users who choose to isolate their feed under these markers. This organisational capacity, in terms of both topics and choice of who to follow, brings together a powerful gathering. It connects interlocutors as the Twitter-gadget gathers in the way of a digital thing – or at least offers the possibility of such at the same time as it tends to disperse and fragment attention.

Gathering is possible because the communication happens in such a way that the expressed thoughts of others can circulate and mutate – loop – in observable forms. For example, loops form in the process of replying, and in the designation of favourites, as in chains of retweets and modified-retweets and so on. As a result the Twitter feeds of clusters of accounts inevitably start to show a regularity in terms of who tweets and who follows. Given the tendency of accounts to focus on certain issues, and for users with an interest in those issues to likewise follow each other, we have groups of accounts/individuals intersecting, retweeting and commenting on each other – forming clusters of shared opinion that are carried on though time. These then form part of the memories of those involved, laying down somatic markers as well as leaving a trace in the archives of Twitter itself.

The Twitter-gadget also entails the possibility that responses may go beyond a purely linguistic or discursive interlocution towards resonance and what can be described as gadget protention. There is a moment of knowing what others will think before they think it, what they will say before they say it, and what they will do before they do it.

My claim is that the consistent use of hashtags, iteration, retweeting and clustering, and other modes of on-the-fly classification, lay down layers of memory over time, producing emotional triggers. These are built up and shared by networks of interlocutors, becoming embedded as background assumptions, or somatic markers, in their brains. This can then support a significant degree of collective emergent behaviour without the risk of a mob effect. This is because the gadget allows for higher levels of looping and feeding of shared social endograms. In short, through Twitter we can get inside each other's heads, our voices can loop into each other's in such a way that we move closer towards a 'thinking' assembly.

Here the patterns across the whole network echo to create synchronous moments and resonance, in some ways analogous to the coordination of the different regions of the brain via the long axons that link together the different cerebral regions and that are vital to consciousness. These extended complex live clusters scale up sufficiently to create a consciousness of shared experiences, of empathy and mutual awareness, such that parallel activities and thoughts emerge into a common clearing with a coherent sharing of ideas and actions. The most observable instance of this comes through the use of hashtags that scale up to break out of smaller clusters of mutual followers and start trending across Twitter. Often hashtags are banal, or generated from marketing or other campaigns, however some are far more profound. For example, the #MeToo hashtag developed from one initial tweet into a social movement in very short time (Khomami 2017). The hashtag functioned in the realm of consciousness in at least two ways. Firstly, it worked to activate knowledge of a shared set of experiences, to generate self-consciousness via iteration and interaction, either through recognition and empathy, or actual participation and contribution, thus extending the cluster further. Secondly, the hashtag created a form of political consciousness, in a very startling form, by gathering a multiplicity of private sufferings into a common realm or clearing – making them available to the social endograms of many interlocutors. This political consciousness then mutated quickly into a number of actions, including the exposure of offenders, but also further actions to create broader media awareness as well as physical protests and demonstrations. The speed of the impact was unquestionable.

This is similar to the way Lefebvre talks about 'the joint efforts of many thinking individuals, in a progressive expansion of conscious-ness' (1968: 48), but here the emergence can occur quickly and with relative ease, because the shared thinking is built around a dyadic and subsequently hyperdyadic spread – the spread of effects from specific social and friendship networks to the wider network. Twitter's capacity for such spread, combined with iteration and variation as feedback, draws together a group dynamic built on already existing triggers and sympathies and emotional as well as rational ties. Twitter then presents a powerful prosthesis; its scale-free topology, its operation at great speed, its collective indexing and its wide dissemination of ideas, produce a qualitative shift of capacity. Thus, I suggest this capacity can augment the 'strange loops' of consciousness and indeed provide new capacities in so doing.[4]

Twitter is not a brain, but it doesn't need to be. It does not 'think' in the classic sense; rather it is a host for our brains that amplifies their affordances into something else, providing a space to think in that also enables us to think with each other. The speed of the interaction means that the time of communication moves closer to something like what Manuel Castells (1997) refers to as 'timeless time'; that is, a form of temporality that is not sequential or linear but transversal, in which different time zones become integrated in special configurations at moments of congruence.

Congruence can manifest itself as a collective 'mood', as when it is said that Twitter is 'angry' or 'sad'. Twitter allows for ideas and moods to scale up quickly and exponentially. This is the power-law logic of scale-free networks. Reactions against injustice, or claims for recognition, are combinations of ideas and emotions that are often transmitted though affective connections triggering the somatic markers built up through more long-term relationships and deliberations. Castells has recently argued that precisely these emotions – anger, rage, fear – are often what push words into action, or indeed repress it.

There is a clear analogy here with John Holloway's 'scream' (2002), which he describes as the visceral reaction to injustice, oppression and exploitation, and as the starting point of collective consciousness and action. But beyond this raw reactivity, wherein a group self-awareness emerges, another of Castells' claims becomes relevant, namely that 'By becoming known to the conscious self, feelings are able to manage social behaviour, and ultimately influence decision-making by linking feelings from the past and present in order to anticipate the future by activating the neural networks that associate feelings and events' (2009: 141). Though I would argue that this association of feelings and events is not affective in the full sense because it is layered though recursive meta-cognitive processes: strange loops strike again. The pure subject of thought is, as Hofstadter argues, an illusion. But the material of thinking is not, and in that sense one can certainly make a case for Twitter as a loopiness machine that provides the affordances for a form of Twitter-thinking. This can extend to what I would call phatic

---

4 It can also produce highly problematic mob-like 'Twitter storms' and other such phenomena; this is further explored in Chapter 5 in relation to the concept of 'idiotism'.

reporting; that is, the recounting of low-level activity and thoughts that constitute minor and quotidian aspects of life, but which also tie together followers and followed such that each reinforces the other's perspective. They become familiar to the extent that individual interlocutors have a pre-emptive understanding of what the others would do or think in response to particular ideas, circumstances or actions.

However, there is an attendant danger here that a small number of voices may become dominant and begin to direct the multitude in the manner of a Hobbesian sovereign. This can produce a mob-like effect that confirms the worst suspicions of thinking around crowds. The clustering of connections around popular hubs that increasingly tend towards centralisation and broadcast models is one of the side effects of a scale-free network such as the internet – this is the power-law effect. Unlike in formal democratic systems there are no checks and balances yet in place, no filtering processes or deliberative norms to offset this danger. Whether they can or will be put in place, and by whom for what purpose, remains a live political question. There is certainly an argument for a democratically accountable regulator, or indeed a more radical solution in the way of social media sites becoming collectives or public entities – issues that will be addressed further in Chapter 6.

There is also the opposite danger of the prevention of sufficient upscaling where coherent, loopy and reflective consciousness is disturbed or never occurs. Disturbance is, of course, one of the other major features of gadgets – certainly of commodity gadgets such as smartphones and tablets, which are designed with distraction as a designed affordance. As discussed above, Twitter reproduces this ambiguity and has this pharmacological character. It is capable of developing long chain links and intersubjective loops that reach across the network, resonating and coordinating thoughts and actions. However, as we know, conscious-ness is very susceptible to goal interference, and Twitter can likewise be distracted, its networks broken up. The more fragmented and distracted it becomes, the more the flitting of attention escalates until it increasingly resembles the crude reflexive reception-action cycle of non-conscious brains.

Alongside marketing, advertising and celebrity self-promotion, and other such distractions familiar within the political economy of gadgets, there are more profound and purposive forms of interference. One example is that of interference by 'bots'. Bots are small programs that are designed to respond algorithmically to certain keywords, names, topics and so forth. They are particularly disruptive on platforms like Twitter, where fake accounts operate semi-automatically to create false outrage or to react in the way of 'flak' to particular networks of interlocutors. We can understand the use of bots on Twitter as in many ways a form of purposive goal interference. By inserting them into ongoing long-chain loops, the creators of bots seed disturbance and distraction, a form of cognitive pain that prevents the development of a more advanced gadget consciousness. Bots are the equivalent of noise in a communications system, akin to electrical interference in the brain's synapses, where connections can no longer be made or sustained. However, bots are not the only form of intervention; the associated practice of 'trolling' can also act as interference, but can also be more nuanced, creating its own kind of network and collective intelligence. This would be one example of what I refer to as an idiotic collective, and which will be developed more fully in the next chapter.

This is the major political challenge for Twitter, and of gadgets more broadly: to balance grassroots self-organisation with the need for formal or quasi-formal modes of filtering, deliberation and representation. We also need to keep in mind that Twitter is not a part of the commons; it is a commercially owned and private software platform, as are the majority of such popular social media platforms. The problems of archiving, searching and utilising its full potential have been made clear, for example with the withdrawal of Google's real-time search. This service allowed a keyword search of the Twitter archive but was withdrawn because of copyright infringements, effectively inflicting long-term memory loss on any Twitter 'consciousness'. Here we encounter the need for what has been termed algorithmic accountability.

The contradictions of Twitter as a facilitator of gadget consciousness are thus found in the tension between its foregrounding of individuals – and the consequent quantifying of popularity under the logic of aggregation – and its being a medium necessarily rooted in dialogue and community. Dialectically speaking it is in the synthesis and the overcoming of interference that the consciousness of Twitter occurs; this is where Twitter 'thinks'.

To summarise, Twitter's combination of short-form micro-blogging – offering succinct but fully formed propositions, observations, imperatives and so forth – with scale-free connectivity at great speed, and with the capacity to cluster and disseminate bounded frames of discourse, produces a qualitative shift of capacity. This can be equated with the idea of a phase shift. A phase shift occurs when a particular substance or entity reaches a certain stage at which there is a shift to another kind of state, for example when water turns to ice, or when a crowd turns from a collection of individuals into a coordinated group with a singular purpose. Thinking and feeling brains and bodies, when intensively connected and communicating, reverse the established lines of communication from the centre to the periphery; they thereby allow for organisation as a process of deliberation and affect layered through intermediation. Here we move to the political level because this shift, at the very least, produces powerful large-scale responses that make it increasingly difficult for the established political class to govern as they would like, that is with a free hand. This is what Noam Chomsky would call – for this political class, and with heavy dose of irony – a crisis of democracy.

5

Gadget Action

This chapter is about the role and potential of digitally augmented collective action, in particular via the process that I have defined as gadget consciousness. Gadget consciousness is highly pertinent to the formation of a collective's awareness of its togetherness, purpose and aim. It is akin to what I referred to as 'gadget class consciousness' in the previous chapter, which is a step beyond the process of collective volition, will and thought, and involves the need for collective self-definition, the self-consciousness of a group or class with an attendant notion of interests and political motive. As such, the idea of gadget action turns on the question 'who are "we"?', or even more potently on the attendant question 'what are "we" *for*?' This entails developing further the question explored throughout this book; namely, how does, or can, the 'we' think together with a sufficient degree of 'collective intelligence' to make the 'we' an active entity with agency? In posing the question in the first instance I am asking how, in the age of social media, we think, decide and act collectively, and how this collectivity is translated into action.

My overarching thesis is that we can see a spectrum of tendencies in collective action that broadly divides into what I will define as *resonant* and *idiotic* consciousness – a division that also equates to the use of gadgets in the mode of things or of objects, and in different forms of active gadget consciousness. Where we are disposed towards our gadgets as objects they become a means of gratification, sources of access to the commodity form in the mode of reification of self and others. This is evidently *idiotic*. Where we are disposed towards our gadgets as things, as a source of gathering, of shared experiences and moments of significance with others, of flourishing and self-expression, then this is evidently *resonant*.

GADGET RESONANCE

In previous chapters I have made the case for resonance within and between brains and gadgets. My starting point was the link between individual and collective consciousness. My presumption follows claims made by Douglas Hofstadter that consciousness emerges from 'strange loops' or 'level-crossing feedback loops'. These are recursive patterns that occur in brains to generate self-awareness, and more broadly, the living construct of self. Of key importance in taking this argument forward is Hofstadter's claim that 'the twisted loop of selfhood trapped inside an inanimate bulb called a "brain" also has causal power', and that this happens 'via a kind of vortex whereby patterns in a brain mirror the brain's mirroring of the world, and eventually mirror themselves, whereupon the vortex of "I" becomes a real causal entity' (2000: xxiv). This causal power is the ability to interpret, create and effect change in a way that goes beyond the purely algorithmic-programmed function of the machine.

Hofstadter further argues that the strange loops that constitute individual consciousness spread beyond single brains, suggesting that different brains are perfectly capable of becoming enmeshed with shared perceptions, thought patterns and so on. In fact, he argues that '[s]ince we all perceive and represent hundreds of other human beings at vastly differing levels of detail and fidelity inside our cranium ... we inevitably mirror, and thus house, a large number of other strange loops inside our head' (2008: 207). These he refers to as 'entwined feedback loops' (210). As such there is a resonance between brains and the production of what I have referred to as a social endogram. It follows from this that certain forms of collective action can emerge from such resonance when there is the emergence of a collective will to do so.

I have argued that the extended feedback loops also entwine with gadgets, and that the distinct character and affordances of gadgets, and whether we orient ourselves towards them as things or objects, inevitably provide a frame of organisation within which our entwined feedback loops must operate. Like any other such relationship described by Hofstadter, when online relationships persist over time they by necessity produce that characteristic element of housing others 'inside our head'. The more intensity the connection has, the stronger will be the loop. Of course, connections that are also active offline – with friends, family members, work colleagues, fellow members of political or social movements – will, again by necessity, have extra dimensions of communication and trust that will intensify the entwinement of the loops.

Resonance then develops and extends though gadgets as things: this is the mode of gathering via gadgets, gathering for a collective purpose that has emerged through the mechanisms of gadget consciousness. There are two distinct dimensions of time at play here: velocity and duration. Velocity – the almost instantaneous exchange of messages – allows for the swift construction of a shared understanding of a situation, which can then be directed towards collective decision making via a shared 'consciousness' of it. With regards to duration, the relationships are not only spontaneous and of the moment but persist over longer periods and generate affective unconscious emergent behaviour, maintained by somatic markers and mnemotechnical memory; such is the background and foundation for the effectiveness of the resonant loop.

The term *resonance* here refers to the resonance between polyphonic or polyvalent elements, and echoes Hofstadter's use of the term; he argues that 'it is a necessary and unavoidable consequence of this set of beliefs that your brain is inhabited to varying extents by other I's, other souls, the extent of each one depending on the degree to which you faithfully represent, and resonate with, the individual in question (248). Such gadget resonance, as well as being digital, distributed and somewhat immaterial, is also a profoundly material and embodied process, as mobile devices allow us to move and coordinate in real time and space. Gadgets host our brains, bodies and spaces, and these then feed back into each other to create collective thinking and action. The formation of action from these resonances can take multiple forms and emerge in a variety of contexts, with attendant variations in the nature of the actions. I will now pick out a number of ways distinct kinds of collective action articulate with gadget consciousness in order to consider the optimum orientation towards a gadget as thing. I do not claim that these are either exhaustive or definitive; indeed they can only really constitute illustrative snapshots, and as such are heuristics to provide a way of exploring the possibilities and affordances of digitally augmented collective action.

RESONANT ACTION

The Occupy movement flowered in Zuccotti Park, New York City in September 2011, having been sparked by the Indignados movement in Spain and the events of the 'Arab Spring'. Occupy defined itself as 'the 99%'. This moniker became a viral term that spread across the internet and into the material spaces of occupation around the world, becoming a pivotal signifier of identification and solidarity. Vital to its potency was that it invoked a stark split between a well-defined 'us' and 'them'.

The 'we' in this case is evoked and conjured up only in this most blunt of divisions. 'We' are the victims of neoliberalism, unfairness, exclusion and exploitation. This is a moment of expression of the aforementioned 'scream' (Holloway: 2002). For Occupy the scream was one of outrage at the dominance of the 1%, and at the same moment the recognition of membership in the 99%. In this moment the 1% are framed as the perpetrators of injustice, and as such they are the enemy. The 99% is the signifier under which experiences can be connected and amplified through social media, creating a looping resonance between the multiple persons involved, without them having to share precise agendas or specific concrete proposals. This is augmented by stories, pictures, videos, tweets, blogs and of course physical connections in the squares.

Occupy relied widely on chains of social media and gadgets. Most Occupy camps had media centres that acted as clearing houses for multiple communications and were interconnected with the global multiplicities of participants via their gadgets. The fabric of the multiplicity can be understood in terms of concentric rings of gadget consciousness; these were most intense in the squares and camps themselves but rippled out around the globe. As Anastasia Kavada observed in her study of Occupy, 'social media helped in diffusing the clear distinction between the inside and the outside of the movement'; 'social media followers formed an outer ring while the inner ring included activists who were participating regularly in the physical occupations' (2015: 878).

The forms of consensus-building in the squares can also be seen as outcrops of such gadget consciousness. An example of this was the use of the human microphone: those physically present in the squares, who were within earshot of a speaker, would repeat what the speaker had said, causing a ripple of repetition back through the larger crowd allowing all to hear without the aid of amplification. This process would happen in reverse when members of the crowd asked questions or offered comments; thus a form of crowd-based communication was inaugurated that echoed the digital forms of social media, in particular comment threads and the patterns of retweeting on Twitter.

A variation on this pattern of resonance occurred as part of the Hong Kong 'Umbrella Movement' in September 2014, which was one of the last actions associated with Occupy. It was known as the Umbrella Movement because of the use of umbrellas and wet towels to defend against police tear gas, and as such the umbrella became a potent symbol which spread around the world. The protest was directed against restrictions in the democratic mandate of the Hong Kong executive. It followed some of the patterns of the Occupy moment – an identified group as the target of protest, and resistance against a perceived injustice – in this case not so much the economic injustice of the 99% as the political exclusion of the 99.9%. While the protests were certainly youth-oriented, the key factor for participation was not age but social media use. This is indicated in a study of the movement by Lee, So and Leung, who observe that social media 'is demonstrated to have a consistent impact on people's support for the protest movement and anti-establishment sentiment' (2015: 368). This study concludes with the claim that the use of social media helped in 'building up dissent, cultivating a common consciousness and identity for a cause, and disseminating dissenting views and insurgent activities across the city and to the wider world' (372).

Also of particular interest in this movement was its use of a specific smartphone app, FireChat. FireChat is distinct from a number of other text and image sharing apps in that it does not rely on Wi-Fi or a mobile data connection but on ad hoc networks, or meshworks, built from the Bluetooth connections of smartphones. This works on a peer-to-peer basis, connecting individual phones together to form local networks capable of multiple peer-to-peer communications and data transfers. It works regardless of any further internet connectivity – though once any one user in the mesh is connected to the internet then this connection can be shared across the mesh. FireChat is built on a meshwork technology known as 'MeshKit', which claims the ability to 'Share and distribute content to users who are not online. Source real-time news from users in the field, even when users or reporters are not connected', a capacity which, the designers claim, allows users to 'Send out early warning and recovery advisories before, during, and after a natural disaster. Reach more people regardless of cellular infrastructure or damage. Collect citizen information and increase community resilience' (Meshkit: n.d.).

The use of this technology in Hong Kong allowed the protest to continue and made coherent communication and organisation possible even when internet connectivity was denied. As was reported at the time, 'FireChat got more than 100,000 new sign-ups in Hong Kong in under 24 hours; it has registered 800,000 chat sessions since' (Bland 2014). The uses included basic coordination of resources on the ground, sharing requests for water or food, as well as live organisation of protests, circulation of information about the activities of the authorities and so forth. The nature of the meshwork created by FireChat means that as more and more people join the more strengthened and resilient it becomes. This gives it a particular utility in dense urban settings, and so it combines very well with crowd-based protests focused on the disruption and control of urban spaces. The patterns of resonance are clear here, and the place of the gadget as a focal point for gathering works in both the Heideggerian and the literal sense of the word.

This gadget assemblage of a smartphone app and Bluetooth connectivity, held in the hands of protesters and locked in step with a multitude of gadget brains, offered an intensive moment of gadget consciousness. The clear vein of collective intention running through the protest was also very strong in this case. The use of the app can be said to have generated multiple loops that captured a shared awareness of the situation in the situation, creating a localised social endogram which enabled the sense and feeling of the 'we' to thrive, thereby supporting effective collective decision making. The consciousness also spread further, both regionally and globally, if in a somewhat weaker form, as the images and symbols of the protests proliferated via other social media, including dissemination via a Twitter hashtag and then into mainstream media broadcasts.

The protests were eventually overpowered as weaknesses in the meshwork were identified; as an open technology it was vulnerable to incursion and manipulation by the authorities. Since then, however, the app developers have sought to strengthen security by introducing end-to-end encryption. This kind of use of the app has been repeated in a number of other cases, for example in the Catalonian independence protests and referendum, which had to be organised against a repressive state machinery. As one participant reported: 'Some of us were already using FireChat to communicate in the crowd because of the slow internet. When the [referendum] organisers told us to turn off our data, we could still share information' (Forbes 2017). It was also used at protests at the University of Hyderabad in India (The Times of India 2016).

The processes just described, both digital and physical, allow for the self-recognition of a 'we', and bring into being a frame within which other concerns and shared perspectives can gather – further expanding the resonant loop into a global overlapping weak gadget consciousness, with more or less intensified local variations. However, it became clear that what these movements lacked was a longer-term durational form. As the short-lived peak moment of Occupy passed, the movement transformed into something more akin to a fragmented number of inter-locking affinity groups, as instanced by publications such as the *London Occupied Times* or the development of a group such as UK Uncut.

However, we have also seen an evolution into more traditional political forms such as, for example, Podemos in Spain and Momentum in the UK. Here we can see the potential development of such resonant collectives into new manifestations of the party form. One such example is X-Net, an organisation set up specifically to leverage the networking power of the digital in the party form but without reverting to top-down leadership. It consists of what one of its founders and key spokespersons, Simona Levi, calls 'widgets'. These are designed around specific goals and clusters of people and technology, an idea which echoes very clearly the notion of a dispositif, and indeed that of a gadget assemblage. X-Net was set up with the remit to expose and challenge corruption, in the wake of the 2008 financial crisis and the ensuing movement of the squares which had foregrounded the idea of a fundamentally corrupt system. Indeed X-Net is also critical of Podemos as an explicitly populist movement that pushed its leader, Pablo Iglesias, into becoming a cult figure. X-Net is a cluster of technologies and a movement that operates on different fronts. Simona Levi tells us that it 'invites leakers to pass on documents that provide evidence of possible corruption, inspired by the WikiLeaks site. But X-Net has gone further, working not just through activism but also in Spanish politics and in the courts'. It also refers to itself as 'a peaceful guerrilla movement'. Levi argues that 'We're the next step forward after WikiLeaks' (AP News 2014).

The Partido X or X-Party, is one element of the movement, with an aim to instil the collaborative and open approach fermented in the 15m movement into formal politics, that is to reshape politics into a far more collaborative and open commons using networking protocols and devices. Two main goals are to establish 'Citizens' Open Lists' and a 'Citizens' Network Lobby'. The X-Party acts as a kind of legal funnel for the grassroots-organised network lobby.

The organisation is conceived both technically and communicatively. There is a technical element in the sense of using the intersections of the internet, mailing lists, web and mobile platforms for sharing (indeed X-Net are even currently developing a payment system to circumvent the banks). They explain that 'we consider that the Internet is an historical era in which we have the obligation as citizens to take advantage of its possibilities to impact, through technopolitics and hacktivism, our society to make it truly democratic, to empower ourselves and to desin-termediate the access knowledge' (X-Net, n.d.).

Thus, X-net and their like can be understood in terms of being a 'gadget party', but one recognising that purely self-organising and horizontal organisations springing out of distributed networks don't really exist, and hence there is a need for face-to-face meetings, for movement towards consensus as well as multiplicity and difference. In that regard there is a possible future in which resonant action can be seen to evolve as a concatenation of gadget + party, with the gadget conceived as a thing and the party as a form emerging from the gathering, which as such includes releasement as the attitude towards gadgets.

While these particular forms are very different, and specific to the politics of their national and cultural contexts, they do have their roots in the resonance that is associated with gadget consciousness. So, while Occupy fizzled out, it has inaugurated a trajectory. The question of what the continuing 'we' constitutes is thus always ongoing and troublesome, in particular in the moment when their agonistic attitude, their identification through resistance to an actually existing authority, requires its own binding logic and set of concrete aims; here again fragmentation and loss of solidarity is a highly likely danger.

We can see another approach to this problem in the attempt to create a form of resonant solidarity that is far more fluid, and that does not appeal to party as faction or set. This is the approach adopted by Anonymous as a political actor. Anonymous is a loose association of hackers and pranksters, operating largely online but also spilling out into street action, most notably in their 'project Chanology' in which the group famously took on the Church of Scientology with online trolling attacks as well as street protests. In the latter, the Anonymous moniker, adopted on the bulletin boards that were its birth place, manifested on the streets with the use of the Guy Fawkes mask. Project Chanology focused on the Church of Scientology because it was understood to stand for a set of values antithetical to those of Anonymous, especially in its attitude towards information. Scientology is proprietorial, closed, prone to the cultish worship of individuals, dictatorial and utterly without humour or self-deprecation (Beyer 2014; Coleman 2015). Anonymous, in sharp contrast, is for complete openness of information, playfulness and mischievous disregard for rules and social homilies. Its activism is not an attempt to coordinate a multiplicity as an overlapping tangle of differences, but to constitute a 'we' through a universalising appeal. This is not an appeal to the universal in the sense of an abstract transcendental human right, but an appeal to a concrete ideal type under which identity is given up to total collectivity for a temporary period. This is close to what Alain Badiou defines as the generic; that is, 'the identity of no-identity, the identity which is beyond all identities' (2006).

Anonymous uses the powerful image of a huge crowd, rendered indistinguishable behind their Guy Fawkes masks, which is taken from the film *V for Vendetta*. The choice of the mask to represent, or rather conceal, the identity of the hackers and activists that constitute Anonymous is a powerful metaphor for the generic. The resonance of the 'legion' behind the identical masks produces a universalising moment wherein the 'anons' stand in for a general, but concrete, condition, we might even say for the hacker class as the generic universal, as defined by McKenzie Wark (2004). There is no standard bourgeois universalising claim (justice, emancipation and so forth), rather the connection between the members of the collective is in the mode of the generic, what can be described as, again, a form of resonance. The digital roots of Anonymous, in the chat rooms and bulletin boards of 4chan, allow for this blending of distinct individualism with resonant styles and most importantly humour. Humour is one of the loopiest, self-referential of things, requiring recognition of a world-view, and shared meanings, genres and references; in the case of Anonymous it facilitates a distinct entwinement and a recognition of an outsider identity that has been excluded from mainstream culture and politics.

Such a prankster sensibility contains elements of resonant action, but adds to this an attempt to absorb the excluded, to become generic in the sense of representing that which cannot be represented, which is beyond the current political frame. The mask represents a resonance characterised as much by style as substance, and the enemy is a shifting target, identified by their failure to get the joke. However, again, the problem of effective change and persistence emerges, of generating a collective will that is capable of sustaining a broader political impact. There is also the distinct and now ever-increasing danger of what one can call the 'forking' problem: the forking of the alt-right from this movement – as the idiot form that can spring from gadget consciousness – to which I will now turn.

GADGET IDIOCY

I use the term idiocy in relation to gadgets as a way of naming a specific orientation towards gadgets as objects, which also entails a particular subjection to them in their commodity form. I take the term 'idiocy' from Neal Curtis' concept of idiotism (2013). The term comes from the Greek *idios*, or private, and relates to *idiotes* or private persons, from which our word 'idiot' derives. While 'idiot' entails pejorative connotations, with its reference to the untrained or lowly and unskilled, Curtis argues we need to recover the primary sense from the Greek (2013: 14). The link with the ancient Greek connotation of 'private person' suggests someone dislocated from public life, not contributing to the common good, who does not 'act', in Hannah Arendt's terms, as a political being. These days 'idiot' has generally been reduced to only the pejorative connotation of stupid, but recovering the sense of the idiot as a private, dislocated, self-serving subject is very helpful as a compression of a set of values typical of the market ideology of neoliberalism. Today's neoliberal reality (what Mark Fisher (2009) termed 'capitalist realism') means that all members of a community are subjects of forced privatisation. This entails being both pushed out of the public realm – an increasingly maligned and suspect entity – but also paradoxically being permanently connected to it. This atmospheric connectedness is another feature of social media and digital communication, one that is highlighted in 'cyber-pessimistic' viewpoints that see such technology as fundamentally alienating – a point of view that normally exists in the absence of any balancing perspective. We can see such hyper-connectivity as one feature of technology operating in the interest of capital, as an attempt to subsume social life into deadened general intellect. But this attempt at subsumption is a contingent and not inevitable feature of technology; it is the social and cultural manifestation of the ideology of idiocy.

The characteristic feature of neoliberalism is its focus on individual interests: property as the primary measure of rights, including the body as private property, and success in the form of profit-making as the driver of all human motivations. These are rendered as absolute natural values. We can see these values present, even if inadvertently, in a number of theorisations of digital collectively. For example, the proposition that digital collective intelligence can emerge simply through the aggregation of individual preferences and decisions is proffered in the popular notions of the *Wisdom of Crowds* (Surowiecki 2004), *Smart Mobs* (Rheingold 2002) or *Cognitive Surplus* (Shirky 2010). All of these texts imagine that we are miraculously smarter together when networked by digital communications, but while still individually making decisions on the basis of our own aspirations. This is a conceptualisation that envisions the underpinning mechanism of society in terms of interactions between self-interested individuals, that is, in terms of markets and their 'invisible hand'. As such these views are rooted in a view of the human being as inescapably singular and ontologically essential, an understanding central to both capitalism and idiotism.

The use of the terms mob and crowd are suggestive in these texts. While they are not meant pejoratively they carry the implication of seriality, of members of a crowd operating in parallel, side by side, but not together with each other. Another common analogy applied in such literature is to the swarm – and this is indeed what I believe is meant by these frameworks and reveals their underpinning ideological assumption. Curtis associates this idea of idiotism with Heidegger's concept of the 'they', *Das Man*. This is not, as Heidegger says, 'something like a "universal subject" which a plurality of subjects have hovering above them' (1962: 166). It is rather a disappearance of the individual Dasein:

> every Other is like the next. This Being with one another dissolves one's own Dasein completely into the kind of Being of 'the Others', in such a way, indeed, that the Others, as distinguishable and explicit, vanish more and more. In this inconspicuousness and unascertainability, the real dictatorship of the 'they' is unfolded. (165)[5]

In the context of gadget consciousness this is not simply about the use of devices for a particular 'idiotic' purpose over and above an otherwise neutral infrastructure, but about the way the political economy of gadgets manifests in material terms within the form and context of the gadget and how it is designed and employed – gadget idiocy is structural as well as performative. This also chimes with Stiegler's view, when he tells us that 'Systemic stupidity is engendered by *generalised proletarianization*'

---

5 It is again worth offering a brief reflection on the use of Heideggerian terms here. There is a danger that drawing on a notion such as *Das Man* leads to an elitist position and a delegitimating of collectives and crowds, and thus to an anti-democratic politics. This is a real concern, but I defend the usefulness of the concept as it is specifically employed here, in a somewhat loose way, to highlight a distinction between different forms of collective that enables a defence of what I refer to as resonant collectivity. Resonant collectives are therefore separate from the 'they' conjured through idiotism. Discounting the possibility of *Das Man* would cut off a line of critique towards particular political configurations – fascism, authoritarian forms of populism and so forth – that undermine the progressive potential of collective consciousness.

(2013: 22), and as such 'the pharmacological question haunts planetary consciousness and the planetary unconscious' (4).

We can thus see the rule of idiocy as taking on a hegemonic character. For example, Jean Baudrillard, who long predicted and explored the tendencies and effects of objects and the digital, saw the emergence of digital networks as effecting a fundamental shift in power relations in line with a move from domination to hegemony, which he mapped out in his late essay of the same name (Baudrillard 2010). Baudrillard sees a hegemonic situation in 'the reality of networks, of the virtual and total exchange where there are no longer dominators or dominated' (33). He links this unambiguously with computerisation, pointing out that hegemony refers to consensus rather than servitude and as such 'brings us back to the literal meaning of the word "cybernetic" (*Kuber-netike*, the art of governing)' (34). This certainly chimes with the mode of idiotism, wherein the hegemonic position is one of ready consent to the neoliberal offer, personal data for services, free labour for sociability, attention for identity. The idiotic variation of this is the ready acceptance of one's primary relations as horizontal: with the platform; with the search algorithm; with the narcissistic mirror of self that captures desire in the circular cybernetic loop (not the recursive differing strange loop). Indeed, Baudrillard offers a very fruitful thought in the context of gadget consciousness when he tells us that '[w]e could compare hegemony to the brain, which is its biological equivalent. Like the brain, which subordinates every other function, the central computer assumes the hegemonic hold of a global power and can therefore serve as an image of our present political situation' (35).

Baudrillard is not alone in this understanding; indeed one of the most common themes in critical theory of the digital is this perspective of encroaching capture, whether in the form of Giles Deleuze's 'Societies of Control', Tiqqun's 'Cybernetic Hypothesis', the Invisible Committee's entreaty to 'Fuck Off Google', Alex Galloway's 'Protocol', Geert Lovink's 'Social Media Abyss', or the notion of 'evil media'. In Matthew Fuller and Andrew Goffey's book of the same name (2012), they posit an unconventional understanding of 'evil' as the operations of media in 'grey' areas underneath human actions. Fuller and Goffey claim that, '[b]y grey media, we mean things such as databases, group-work software, project-planning methods, media forms, and technologies that are operative far from the more visible churn of messages about consumers, empowerment, or the questionable wisdom of the information economy'

(1). It is grey media that keeps neoliberalism alive: 'The transparency of the facilitation of activity that is produced when devices, practices, protocols and procedures, gadgets and applications, mesh and synchronize simultaneously creates vast black-boxed or obscurely greyed-out zones' (13). It is precisely such greyed-out zones that operate to capture general intellect as a form of dead labour, the accumulation of informational capital, the development of algorithmic control that works to pre-empt and direct not only searches but policing, political dissent and collective action (Elmer 2008; Grusin 2010).

As mentioned previously, Franco 'Bifo' Berardi has extended this understanding further into the neurological when he talks of a 'technomaya', wherein we have a media that is 'directly acting on the mind, so that the spell of the media-sphere has wrapped itself around the psychosphere' (2014: 6). It is the attention economy that now breaks up our ability to think freely, and we find ourselves in 'a cage for future action and for future forms of life' (7). Berardi's solution to this is to be found 'in those places of the unconscious where the multilayered spell of semiocapital is ripped apart in order for a creative unconscious to resurface' (7).

In idiotism there is, to put it in terms of the previous discussions, an absence of loopiness and resonance. The form of the association generated by resonance is defined by the nature of the objective and the degree to which acts are shared and considered, but also by the context of the coming to consciousness. With the idiotic we see a focus on blinkered interests and self-gratification, even if at times this is mitigated by an aggregation of such limited self-interests. These are intensely privatised, oriented towards an individual or closed group, hedonic, self-oriented and as such readily captured by an atomised consumerist sensibility.

The idea of a collective that is tinged by idiocy is at first glance contradictory, in that idiotism bends the sense of the collective away from its core meaning, of shared and mutually supportive groups, towards a sense of fragmented and individuated atoms. In that sense the accumulation of such atoms in collective-like forms is better described as an aggregation, even if the appearance of such forms can look and behave like a collective. While it is important to remain mindful of this tension, what I am describing is a spectrum rather than a specific measure – an intensive rather than an extensive scale – and as such some actions, indeed most actions and forms of gadget consciousness, mingle these two forms to some degree.

FROM RESONANT TO IDIOTIC ACTION

Student protests

The first example I will look at is just such a mixed and ambiguous case: a gathering with resonant practices but one in which elements of idiocy are present, but not hegemonic. This was the series of protests against the rise in tuition fees in English universities.

'F**k Fees' read the placards of the students and educators who marched past the palace of Westminster, many demonstrating for the first time. The UK student protests of 10 November 2010 revealed the vehemence and anger of young people, and this spilled over into disorder at the Conservative Party headquarters in Milbank Tower, where a group of students stormed the building and occupied the roof. This led to one student throwing a fire extinguisher over the parapet of the roof – an action that predictably captured the newspaper headlines the following day. The *Daily Mail*, being nothing if not predictable in its ire, referred to the 'hijacking of a very middle-class protest', blaming a core of 'anarchists' who 'whipped up a mix of middle-class students and younger college and school pupils into a frenzy' (Gill 2010). In one respect the *Mail* was right about the middle-class element of the protests, though 'middle class' is not quite the correct term to apply to the action, and 'consumer-oriented' would be more accurate. The underpinning logic of the protest was one of refusing a 'bad deal'. The support received from across the student body included middle-class parents as well as younger teenagers protesting the withdrawal of the EMA (educational maintenance allowance). A neoliberal market logic was therefore at the heart of the situation: what was sought was not a general principle of equality or of universal access but an ethic of self-care and an attendant resistance to perceived poor value for money.

We can include this example in the context of digitally augmented collectives, given that in the moments of action a multiple, multidirectional, highly mobile and highly communicative group, creating new relations to space, time and each other, was formed. However, what emerged was a political equation balancing the cost of education with the monetary value of a degree in the labour market. The 'we' of this moment thus became more an aggregate of 'I's – and as such far more difficult to form into a longer-term resonant collective. This is not to denigrate the cause and its legitimacy, but to see its divorce from a wider field of politics and social relations manifest in its idiotic tendencies. Because these involved short-term and atomised relations, entwinement did not occur; social media foregrounded the elements of velocity without the tempering power of shared consciousness that requires duration. This does not detract from the legitimacy of the protest but reflects its composition by students too captured by their assigned role as consumers, becoming aggregated under specific local circumstances. As such we can refer to this as a quasi-resonant action, in as much as it still operated within the prevailing political situation, which tends towards action that does not challenge the market form as such. There is still emancipatory potential here – the action was not purely self-interested, and there was a shared purpose – but its idiotic component rendered it easily isolated and defeated.

The emancipatory potential is present in at least two forms; firstly, the protests revealed a nascent development of gadget consciousness in the attempt to develop specific technologies to support collective action and build on the affordances of mobile communication. Specifically, such affordances were indicated by the development of the 'Sukey' app by a number of activists. Sukey was an attempt at live geo-mapping to track protest action and feed back information into the crowd's actions: 'Sukey brings together in-house code (fuelled by many late nights), resources like Google Maps and open-source software like SwiftRiver' (www.sukey.org). Sukey was an attempt to gather people, code and data – to create an opening, a horizon of possibility. In fact, the SwiftRiver software it drew on itself explicitly employs the notion of gathering: 'Gathering and filtering information from a variety of channels (RSS, Email, SMS, Twitter. Etc.) Drawing insights from the collected information.' Accordingly, the app had the potential to orient users towards their gadgets as 'things', in particular at the point that the app/phone assemblage hits its phase shift, becoming the main affordance of the gadget at the time of use. Thus, the gadget, here constituted as phone + Sukey, gathers correlates of persons (or one might say dividuals) in a way that maintains a form of releasement. That is, a disposition towards the gadget as working through it towards the others gathered by it, drawing on it but not being directed by it – and importantly placing it aside when appropriate.

Unfortunately, both SwiftRiver and Sukey have ceased development, but they can be seen in a media archaeological light as what I would describe as retroactive vapourware, that is, they provide an inkling of the possible via an understanding of an historical path not developed to its full maturity. The failures of Sukey were as much to do with the political economy of digital as anything else; it simply lacked commercial potential. As such, while Sukey as a platform was partially effective, in some ways it's more helpful to think of it as an imaginary app – it offers an alternative conception of how we can interact with the present, even cycling towards a new prefigurative technology, but one that needs to be conceived as part of a more general political anti-capitalist movement.

The second emancipatory element in these protests was clear in the politicisation of a generation, and their own increasing use of technology in new and innovative ways. Thus, the movement evolved a number of offshoots that are resonant collectives with more duration than the quasi-resonance of the protest movements themselves. For example, Novara Media is an organisation founded by Aaron Bastani and operated by a group of journalists, commentators and activists who first came to political awareness in the 2010/11 period. Starting as a weekly radio programme on London's – coincidentally but appropriately named – Resonance FM, Novara has evolved into a more comprehensive media operation, run on a shoestring budget of donations. They have put together their own production facilities and distribute material through their own website, but also undertake a skilful and deft use of commercial social media platforms. The Novara YouTube channel regularly garners many thousands of views for its commentaries and interviews, and as of mid 2018 it has over 16,000 subscribers to its channel. Aaron Bastani's interview with David Harvey in late 2017 had 7,000 views by mid 2018 on YouTube alone. There are also a number of regular contributors and sympathisers who form part of an associated alternative media ecology, including Owen Jones and the former BBC and Channel 4 journalist Paul Mason. A debate held by Novara between Harvey, Mason, Bastani and others on 'Post-Capitalism and Technology' had almost 17,000 views between September 2017 and May 2018.

This cluster of activists and journalists has done more than merely present videos and articles online; they have increasingly leveraged the use of technology to support political change via formal politics, in particular through the Momentum movement within the Labour Party. This proved highly effective during the 2017 UK general election, con-tributing to the unexpected surge of support for the Jeremy Corbyn-led Labour Party.

London riots

Going significantly further along the resonant-idiotic spectrum we can find idiotic tendencies starting to prevail and moving towards a hegemonic position. One such instance was the London riots of 2011, which originally began with the police shooting of Mark Duggan in Tottenham, London on 4 August. It was believed at the time by Duggan's family and members of the community that this was an unjustified and unlawful killing, and it provoked several days of protests and increasing disorder.

One well-reported call to action came via Blackberry Messenger: 'Fuck da feds, bring your ballys and your bags trollys, cars vans, hammers the lot!!' (Ball and Brown 2011). This was the first of a number of viral messages that circulated, encouraging rioting that was focused on looting, in particular the looting of shops carrying high-end consumer goods, phones, trainers and other high-value items. The court system and government did not hesitate in reducing the rioters to the clichéd figure of the terrifying crowd, the mob, the mindless masses, and the court system duly handed out highly draconian sentences to set a punitive example.

Yet we can still see this 'mob' exhibiting many of the characteristics of a collective with some 'smart mob' tendencies. While the 'we' articulated here was not a reflexive, considered or progressive one, it can be said to have existed in so far as there was a shared interest in aims, and a communicative process that produced a 'swarm' like effect via the use of messaging services and social media. This was not a mindless mob; indeed it was a mob most mindful to share opportunities, and to identify and outwit its enemies (the police). But, akin to the previous example, we can see this as a more extreme and 'pure' expression of frustrated consumerist desire, not a reflective or overtly political reaction, but one that extends the coordinates of neoliberalism. As Slavoj Žižek remarked on the logic of the riots: 'there is no ideological justification, it is the reaction of people who are caught into the predominant ideology, but have no ways to realise what this ideology demands of them; so it is a kind of acting out within this ideological space of consumerism' (Fiennes 2012). Nevertheless, in so far as the riots went beyond the exchange relation, we can find some disruptive challenge present in them. However, the challenge was so inflected with idiotism that the effect was bereft of almost any political content, and as such could be readily brushed aside by state power and reabsorbed into the stream of capital accumulation.

Why did the riots fork towards idiotism rather than resonance, given their clearly political and justifiable opening spark? I suggest that it is because they unfolded within the logic of the techno-capitalist dispositif, and as such entailed an orientation towards gadgets as objects which was never overcome. Gadgets were used to challenge the urban space to reveal itself as a shopping mall. They were used to spread information about actions with a primarily acquisitive or hedonic drive. There was a lack of loopiness in the communications which meant the social endogram was not one that could include a sufficiently well-defined 'we' in relation to the wider social and specifically class position. The lack of loopiness meant there was a lack of gadget class consciousness that could conceive or orient itself as a collective: recursion did not become reflexivity. The pre-existing social dynamic and ideological framework could not be overcome by any move towards collective thinking, and there was a failure to achieve collective intention – in that sense what remained was an aggregate of limited individual and isolated drives. This is not to say that gadgets in any way caused the riots, but that the specific gadget ensemble present allowed them to develop in a specific direction under the force of drive within the dominant ideological framework. Idiotism plus gadgets equals gadget false-consciousness.

The idiotic form was also mirrored in an inverted version of this: the action that became known by the Twitter hashtag #riotcleanup. This was an attempt to 'counterbalance' the destructive forces of the mob by repairing the damage done in specific neighbourhoods after the riot. The riot clean-up was inaugurated by two individual Twitter users, Dan Thompson and Sophie Collard, and soon went viral. It was picked up and lauded by the press, who used it to illustrate that social media can, after all, do good. The BBC reported the celebrity chef Jamie Oliver, whose Birmingham restaurant had been attacked by rioters, as saying, 'God bless the communities getting together to sort this out #riotcleanup – people who care about their country!!' (BBC News 2011a). Elsewhere on the BBC the clean-up was reported as a 'fightback' and the brooms presented as a 'symbol of the resistance to the riots' (Castella 2011).

The result of the Twitter clean-up viral message was that the streets of Clapham Junction were overrun by a crowd of largely white middle-class property owners waving brooms aloft in preparation to sweep the streets clean of the undesirables, and of all evidence that any unrest had taken place. 'Where is your broom?' the crowd shouted repeatedly at Boris Johnson (at the time the Conservative Mayor of London). A broom duly appeared and, never shy of an opportunity, Boris addressed the crowd, brush aloft, telling them 'you are the true spirit of the city, you represent this city, not the looters and the thugs' (BBC News 2011b). Explicitly rejecting any kind of 'sociological' explanation Johnson called the riots 'criminality, pure and simple'.

Although presented as a victory for decency and the 'true spirit' of Londoners, the clean-up can be equally, and more accurately, described as a defence of property rights against 'criminal' incursion. This must be the case given Johnson's designation of the riots as 'simple' criminality. The clean-up crew's raising of their brooms was an unintended symbol of the gentrification and economic clearance that has been the pattern in such areas of London. The arrival of rioters from the more deprived outlying areas of London – Croydon in the South and Enfield in the north – presented a powerful return of the repressed. Defence of personal property is not in and of itself wrong or necessarily idiotic, but taken in isolation from the context of the socio-political situation and the spatial and material divisions that fracture London, it manifests as contextually idiotic. There are aspects of resonance here, for example in the use of mobile social media around the #riotcleanup hashtag, which echoes aspects of resonant action. There was also some clustering and looping of perspectives and aims. However, these were based on a limited set of bounded interests, which view the world within the confines of a privatised neoliberal ideology. The riot clean-up was thus an act of ideological struggle, precisely in the sense that Žižek terms it: an acting out to maintain and reimpose the dominant ideology. The clean-up presented a curious mirror image of the riots, an unintended and unconscious manifestation of a class interest in sustaining the status quo and its myriad inequities, fighting to maintain the idiotic privilege of a personal corner.

The tendency to reimpose an existing dominant ideology, to reify the 'other', to present the private interests of a group or individual as the general interest, is the primary and dominant characteristic of the most complete kind of idiotic action.

Alt-Right

An example of an even more virulent form of idiotic action was the ad hoc birth of an online mob after the Boston marathon bombing of 2013. Soon after the news of the bombing broke, Anthony Reed, a graduate student at the University of Louisiana, instigated a thread, or 'subreddit', on the reddit content aggregation and commentary website to track news developments. He quickly began to receive first-hand reports and photographs from the scene of the explosion, as well as a high volume of comments and reflections of the event. The site started functioning as an information sharing point, with only one person acting to filter and moderate the flow of information on the subreddit. Reed suggested that initially this functioned well, as he was able to draw on 'people actually on the ground taking pictures and video ... [and] we were able to determine exactly what had happened with precision and speed' (Barker 2015). Another reddit user, OOPS777, inaugurated a second subreddit the following day, the 'FindBostonBombers' thread. This started out as a well-intentioned attempt to assist the police in tracking down the culprit, by drawing on TV news reports and using the large number of amateur photos and video footage submitted to reddit. One of the moderators, Chris Ryves, was aware of the dangers of vigilante justice and issued a plea, saying 'let's not turn this discussion into something racist'; he wanted to help the Reddit community 'sort through this tragedy as best we could' (Barker 2015).

Users began sifting through the material, attempting to locate the suspect/s. However, this 'rapidly devolved into a farrago of unfounded speculation, amateur detection, crackpotiana and conspiracy theory' (Miller 2015). The thread became awash in overt racism, threats of extreme violence and an almost classic mob sensibility. Claims circulated on reddit then migrated to Twitter, diffusing ever more widely until one of the speculative, and distinctly racialised, stories – accusing a Moroccan high-school runner of being the terrorist – was published in the *New York Post* and discussed on mainstream TV news channels such as NBC and Fox News. The speculation spiralled further with the identification of a named suspect, who had gone missing in the previous month, and who was then further pursued on reddit and Twitter. It later transpired that the individual had committed suicide prior to the marathon taking place. Meanwhile, the family of the missing person had suffered days of speculation and extreme hostility.

What this occurrence clearly indicates is the potential for an online mob to congregate. While further research has argued that such a mob formation is not normal, and that 'the efforts and attitudes in *FindBostonBombers* are not indicative of reddit as a whole', it did happen. The fact that 'the culture of reddit, with its need for evidence, hierarchy information flows, and link structure, is not a good fit for the kind of activity that went on in *FindBostonBombers*' (Potts and Harrison 2013: 148) tells us that it can happen anywhere. Regardless of the 'normalcy' of this specific action on reddit in and of itself, such actions clearly can and do happen. In some ways this action was an inadvertent prototype for much more virulent and purposeful acts of idiotism that have subsequently evolved, for example the 'gamergate' controversy in which female video-game journalists were hounded and attacked with vitriolic insults and threats of harm and death.

Some of these tendencies can be traced back to the same sources that gave birth to the progressive aspects of Anonymous – message boards and online communities such as the /B/Board and 4chan. While Anonymous forked towards the resonant, other factions went in this very different direction. The trolling culture that thrives on offence, racism and misogyny found a fertile ground on these boards and has undertaken a continual process of mutation and expansion, both online and increasingly offline too.

The rise of the 'alternative' right, or 'alt-right' for short, is also associated with this trolling orientation and has been behind the penetration and manipulation of social media by hate groups peddling racism, sexism, white supremacy and other variations of nationalism and xenophobia. The alt-right is a grouping of extreme right-wing self-proclaimed 'outsiders' connected with the rise of 'fake news' and the increasing return of totalitarian political tendencies. It has been linked with the rise of Donald Trump in the USA – his well-known association with Steve Bannon (founder of the alt-right website Breitbart News) and others was a significant issue of debate in the early months of Trump's presidency. The journalist Mike Wendling has studied the alt-right extensively and observes that 'central to the alt-right's conception of itself is that it represents something fundamentally countercultural – activists have compared their movement to punk rock or the hippies of the 1960s. The comparison stems not from shared political values but from the alt-right's claim to "outsider" status' (2018: 8).

The alt-right is widely understood to have its sources in online platforms like 4chan and reddit, and when mixed with other expanding reactionary social forces this creates a potent mixture. In her exploration of the evolution of troll culture and the alt-right, Angela Nagle makes the point that 'It was the image and humour-based culture of the irreverent meme factory of 4Chan and later 8Chan that gave the alt-right its youthful energy, with its transgression and hacker tactics' (2017: 13). She observes that there has been 'a gradual right wing turn in Chan culture centred around the politics board /pol/' (13), and concludes that 'what we call the alt-right is really this collection of lots of separate tendencies that grew semi-independently but which were joined under the banner of a bursting forth of anti-PC cultural politics through the cultural wars of recent years' (19).

The alt-right represents a potent outcrop of the idiotic. In many ways it appears different from the formulation offered previously, in that its adherents, loosely affiliated as they are, reject the kind of economic neo-liberalism and globalism that is obviously associated with the atomised, privatised self of the idiotic hypothesis. However, this self-proclaimed rebel status – aimed against an imagined enemy with many names, be they 'political correctness', 'social justice warriors', 'snowflakes' or some other such moniker – is actually conceivable as pure ideology, railed against a set of straw men. If these targets are genuinely understood as such by the members of the alt-right then I would suggest this is a case of false-consciousness, or at least self-delusion. Given the centrality of gadgets in this configuration this can again be framed as a case of gadget false-consciousness.

Such gadget consciousness is false in that it proclaims a defence of an imaginary individual freedom that is being curtailed, but in reality is a reactionary defence of the privileges of the already powerful. Indeed it is hardly original either, as Wendling observes:

> the argument that the alt-right represents a 'counterculture' comes almost entirely from the movement itself and rings hollow when properly examined ... In actual fact, the alt-right is quite a culturally sterile place – producing a bunch of Photoshopped images ('memes'), tweets, propaganda videos and in-jokes, sure, but very few original songs, bands, films, or other cultural artefacts. (2018: 9)

In other words, this is not a resonance amongst a polyphony but a strange process in which all views become captured or inverted into a monotone singular meaning. The reinforcement and recursions of this form of gadget false-consciousness amount to a reimposition of the status quo and the existing economic, racial and gender hierarchies, but in a more extreme form. Certain gestures towards a critique of neoliberalism by, for example, 'paleoconservatives' and others have proven very thin in the wake of Trump and his circle coming to power. Rather than representing a challenge or a resistance movement, the alt-right and other idiotic manifestations of gadget consciousness are movements of reaction: a fall back on exclusion, essentialism, rigid thought and hierarchy. This is reinforcement without creation, the pathological side of the gadget-brain manifest in political form. It creates not loopiness but rather adherence to a norm through a process of exclusion, othering and identification.[6]

This argument returns us to the question concerning gadgets, of our orientation towards them, and whether it's possible for a gadget to *thing* when so much in our culture and technology conspires against this. I suggest that it can, given the possibilities of *resonance* covered, but that *idiotic action* is within a frame that is defined by *gadget false-consciousness*. Undoubtedly this orientation is encouraged and even driven by affordances designed into the technology, and indeed the kind of algorithms built into many platforms and networks mitigate against gadgets 'thinging', but this represents a broader social and cultural orientation that needs to be pursued collectively and with intention in the context of the materiality of gadgets and political economy. The paradox of having to do this with the self-same gadgets is one of the great challenges we face. The fact that movements do fork indicates that this is not only possible but real – hence the split in the ecology of Anonymous and 4chan into the progressive resonant and the idiotic branches.

GADGET EVENTS

The possibility of going beyond the idiotic can be predicated on the recovery of consciousness and the development of collective will, building on the existent forms of resonant action discussed above. I have argued that we can see resonance, entwined loops and the laying down of somatic markers bonding resonant groups together and distin-

6   There is a possible objection here that I am conveniently labelling all left-leaning collectives 'resonant' and all right-leaning ones 'idiotic'. This is not the case by default; there are certainly right-leaning movements and ideologies that have elements of resonance, for example libertarian movements which emphasise the individual as the primary source of rights and responsibilities, but also recognise a base set of protected individual and social needs and values. Likewise, there can be idiotic movements on the left, not least of which is Stalinist communism. However, I certainly am claiming that the strands of neoliberalism, neo-conservatism and the alt-right discussed here, and which are currently dominant in political discourse, are idiotic. I would also add that right-wing thought, and indeed much of what currently passes for 'centrism', bends towards the idiotic in its valorising of the private over the public and the common.

guishing them from idiot collectives. But so far, arguably, it has been the idiotic path that has been more successful in terms of instigating social change, with the election of Donald Trump in the US and the increasing right-wing populist movement across Europe, including in relation to Brexit. This makes it very clear that the pharmacological character of the gadget is real, and the imperative is to be mindful of this, emphasising the context of digitally augmented action as fundamental. Moving beyond idiocy means recovering a collective, conscious volition as a mode of gathering. I have argued that such a form of collective consciousness is possible, and with the aid of social media platforms it can be expanded to create broader 'resonant' action. In this sense consciousness can be conceived as class consciousness, entailing a shared concept or identification among a specifically bounded group, but signalling the generic; in a political context 'consciousness raising' is a cognate term. So it is that we can act both in and for the extended resonant collectives with which we are entwined. This echoes the definition of class consciousness we find in Georg Lukács, wherein resonant collectives can act on behalf of class interest:

> relating consciousness to the whole of society it becomes possible to infer the thoughts and feelings which men [sic] would have in a particular situation if they were able to assess both it and the interests arising in their impact on immediate action and on the whole structure of society. That is to say, it would be possible to infer the thoughts and feelings appropriate to their objective situation. (1971: 51)

To put it in the terms I have been using, this is to conceive of an unfolding of resonant action in both velocity and duration, one that involves micro-organising, coordination and collectivity, and which results in concrete political outcomes. The point is not to start acting purely 'on behalf' of an imagined class that is justified by some 'big other', in the way of a vanguard to elite party. Rather, the key to this form of gadget class consciousness and action is not to claim more than is warranted in the way of an expanded 'self' as such, but to embrace a principle of sufficiency for the action in the name of the generic, that includes the excluded, the marginalised and the exploited. This action-ability therefore includes collective will – the mobilising of the gadget brain. The potential significance of new forms of social media, which enable real-time communication, lies in their capacity to extend and generalise such actionable consciousness. They allow us to conceive of a 'we' as producing an active general intellect for the common good, one that contributes to the realisation of care, becoming and collectivity – holding back from becoming simply 'they'.

But can we conceive of a form of gadget-augmented action that does even more than this? The forms of resonant action we have so far explored lie broadly within a conception of evolutionary social change, of collectives militating for progressive agendas that enhance the cause of care, becoming and collectivity. Could gadget consciousness instigate a more radical and profound change in terms of freedom from exploitation, improved life chances and emancipatory forms of subjectification? To argue for this, I will introduce the concept of an evental resonant action, which will draw on the philosophy of the event developed by Alain Badiou. This will open up the space to consider the more radical, indeed revolutionary, political possibilities for gadget consciousness.

For Badiou a true event is one that introduces a profound break with the established order. It emerges out of the void, that is from elements that are present but not recognised or represented in the current order or situation. An event will also reveal a truth, but not in the sense of an accurate correspondence to a set of facts – such a 'truth' is merely the banal supporting of the current situation. Rather, the form of truth that the event reveals is one that shatters and upends the status quo precisely because it exposes the hidden that is beyond knowledge, and as such is not operating in the realm of epistemology. Indeed, it is a radical break that requires a commitment ungrounded in anything but the event itself. According to Badiou the truths revealed in events are universal because they relate to what he calls the generic.

The generic set contains the excluded part that is not represented in any of the sets that constitute the recognised situation. The excluded part does not 'fit', and as such it is the source from which events burst – because of this it also undergirds reality (reality in the Lacanian sense of the world framed within our symbolic systems). Thus, Badiou argues that '[t]he correlation between the universal and event is fundamental. Basically, it is clear that the question of political universalism depends entirely on the regime of fidelity or infidelity maintained' (2009: 31). Fidelity is so vital because recognising the event, and acting upon it, is a decision that must be made without the micro-level weighing up of pros and cons within the existing terms of reality; it is rather a matter of throwing oneself into the moment, seeing the truth as entailing a commitment. This is fidelity to the event, not to particular doctrine but to the event itself in all its specificity (such as the French Revolution and so forth).

Events come from outside the state of the situation, and because of this they challenge individuals and groups that are embedded in and framed by the situation. Having one's world shaken by an event can, and indeed should, compel those impacted by it to move beyond the limited subject position provided by bourgeois individualism – the familiar and comfortable condition of occupying an accepted place – towards more radical, new and exceptional collective commitments. The decision produces an 'evental statement': a statement that tries to recognise the event and make it manifest, and as such available to fidelity. Badiou uses the example of the statement 'I love you' in the wake of the event of love. The evental statement 'declares that an undecidable has been decided', and this moment is profoundly important since 'The constituted subject follows in the wake of this declaration, which opens up a possible space for the universal' (2009: 38).

In Badiou's ontology a subject is always more than an individual and subjects are only formed in the act of recognising and acting upon the 'truth' that is revealed by the event. The commitment to the truth revealed in the provocation of the event has the form of 'fidelity', which itself generates the mode of subjectivation in the collective. As such it must be reemphasised that the subject is not an individual as such, and certainly not the bourgeois private individual, but a multiple and active constituent of the world. Because events shake up and reshape the ground of the world, an 'event is a surprise'; '[i]f it were not the case, it would mean that it would have been predictable as a fact, and so would be inscribed in the history of the State, which is a contradiction in terms' (Badiou 2010: 12). The state functions as the totalising horizon of the possible, as the realm which oversees 'a life with neither decision or choice ... whose conventional mediations are the family, work, the homeland, property, religion, customs, and so forth' (11). Events thus break through this horizon. Badiou also relates the event to the notion of 'the exception' (2009: 13) and to the 'Outside'. The aim of developing this concept of the event is to open up choices, to explore the contradictions between different regimes of truth – to 'throw light on the value of exception. The value of the event. The value of the break. And to do this against the continuity of life, against social conservatism' (12). The commitment of fidelity thus means 'to be in the exception, in the sense of the event, to keep one's distance from power, and to accept the consequences of a decision' (13).

Subjectivation is then a form of awakening, an activation as part of a collective. It is the fidelity to an event that contravenes and breaks the dominant power of the monolithic state, that politicises the collective and defines it as an evental situation. As such, '[a]ll resistance is a rupture with what is. And every rupture begins, for those engaged in it, through a rupture with oneself' (2005: 7). This combination of the event, fidelity and subjectivation is useful here because it provides a tool with which to think about the ways that collectivity can be realised as an overtly political force, beyond the kind of democratic negotiation that has been so compromised and captured by neoliberal politics and economics. Events are by definition always revolutionary, or at least proto-revolutionary, as they shatter, or threaten to shatter, the dominant order of things.

The coming together of a subject in fidelity to an event also provides us with a way of thinking about how the resonance of gadget consciousness can be part of such fidelity. The strange loops of gadget consciousness, with their ability to exercise shared processes of thought and action, can also be brought into existence in the wake of events; indeed they provide a capacity for shared fidelity in faster and more widespread formations given the already existing infrastructure. Of course, a key question remains concerning the extent to which digitally networked gadgets are able to support such fidelity in light of their capture within the material and algorithmic constraints of communicative capitalism. That is exemplified in the tendency of social networks and platforms to isolate and turn their users inward, to focus on consumption, self-representation and auto-commodification. This is a question I will return to in the final chapter, but for now it will be helpful to illustrate this point with the example of the Egyptian revolution of 2011.

The uprising in Egypt is the most powerful recent example of an evental situation. The mass that gathered in Tahrir Square, Cairo, on 25 January 2011, was the most pivotal construction of a 'we' in recent political memory. While the crowd were of course gathered in resistance to the oppressive Mubarak regime, they were also gathered as affiliated groups of workers, students and citizens with articulated demands and aims. Rather than being directed merely towards a specific group interest, their aims were societal: political freedom, decent working conditions and respect for basic rights. The gathering in the square resonated in mood and affective transmission, and even more powerfully this also contributed to an intensification of a generic character – the crowd manifested a much more general resonance within the country as a whole. Badiou tells us that the power of this 'consists in an intensification of subjective energy' and in 'the localization of its presence' (2012b: 58). This reflects what he refers to as 'contraction' and allows him to argue that in many ways the uprising in Tahrir was emblematic of the whole people, in as much as it became a 'representation of itself, a metonymy of the overall situation' (58). Such a gathering, although locally focused, produces a 'subjective de-localization of the site'; that is, its immediacy extends towards the universal: '"Tahrir Square" is a site the whole world is listening to' (95).

This idea of Tahrir as a site to which the whole world was listening is justified by the intensely mediated nature of the protests. They were widely covered by global news media, but it was also recognised that mobile communication and social media had played a part in the protests. Indeed, there was much speculation about the extent to which the events could be attributed to social media. There was a crude binary of for and against arguments, between commentators such as Clay Shirky, Malcolm Gladwell and Evgeny Morozov. However, Peter Beaumont, as a journalist on the ground in Egypt, argued that regardless of the abstract debates, 'social media has played a role. For those of us who have covered these events, it has been unavoidable ... a mature and extensive social media environment played a crucial role in organising the uprising against Mubarak' (2011). This is a view that is supported by subsequent research. Zeynep Tufekci explains that for at least one of the protest organisers, emblematic of many others, 'Had it not been for social media leading her to others with similar beliefs before the major uprising, she might never have found and become part of the core group that sparked the movement' (2017: 10). Tufecki quickly unpicks the kind of cynical arguments put forward by Morozov, telling us that 'Most people who become activists start by being exposed to dissident ideas, and people's social networks – which include online and offline interactions', and that 'these provocatively written articles [by Morozov et al.] were often used in the competition for clicks online, and often paired with equally unfounded analyses hyping the internet in simplistic and overblown ways', which meant that 'complex conversation on the role of digital connectivity in dissent was drowned out by vitriol and over-simplification' (10). Despite all of this Tufecki is clear that, 'Thanks to a Facebook page, perhaps for the first time in history, an internet user could click yes on an electronic invitation to a revolution' (27). Whatever the specific levels of cause and effect it is a demonstrable reality that Tahrir Square was heavily populated by people and networked gadgets, and that these networks were interconnected, as well as reaching out into the wider world and back again.

With the geographical focus of the square there was great intensity in the gathering – the density of the connections within the multiplicity of the crowd generated a capacity for expressing a collective will that transcended its time and place. 'Digital connectivity', Tufecki writes, 'had warped time and space, transforming that square I looked at from above, so small yet so vast, into a crossroads of attention and visibility, both interpersonal and interactive' (xxxviii) – an experience shared globally by millions of digital onlookers.

This is the essence of Badiou's claim that this was an event that was reaching for the universal, and is in line with the mediated character of the event. The resonance of the mutual commitments of the people, already somatically marked in the brains of participants through political solidarity, was activated by the Event of Tahrir Square and amplified in the feedback loops between the bodies in the square and the gadgets connected to them. Social media, as well as the television coverage, was being fed back into the focal point of the square, with the polyphony resonating more and more. With each iteration of the entwined loop there was an increasing collective direction and agency: recursion moving towards reflexivity. As with the Umbrella Movement, this allowed for the emergence of a localised social endogram, a contextual picture that included itself within the picture. While the prevalence of gadgets cannot be said to cause anything as such, their ability to gather all the elements together and focus and intensify them constantly enriches the social endogram, feeding the collective intention and contributing to subjectivisation in Badiou's sense. While the response did not reach a tipping point in Hong Kong, it did in Egypt.

Slavoj Žižek, in his book on the theory of the event, talks about Tahrir in terms of a 'pure Event, something that just occurs – it disappears before it even fully appears' (2014: 90). Žižek takes his notion of an event from Badiou, in which the context of its occurrence is itself opened up and reconfigured, changing the very terms of that context. While events might 'just occur', we can speculate that in some instances they might occur in parallel with the kind of phase shift that can be achieved with the intensity of resonance that happens in situations such as Tahrir Square. In the same way that consciousness bootsteps itself though the brain and its extensions, perhaps events do not 'exist' in the same way – but are real none the less, emerging out of the concentrated intentions of an ever more reflexive shared consciousness. We can see that in Tahrir, in conjunction with so much else that was taking place in Egypt, this happened to the degree that an event did take place and shattered the status quo. It is then appropriate to refer to the crowd in Tahrir as undertaking an evental action – such an action cannot be foreseen or forced, but its power lies in pursuing the truth within the event and remaining faithful to it.

The imperative then moves on to become one of sustaining the collective will as fidelity to the event. This, Badiou makes clear, is the hardest part – it is the 'question of *organization*, or the *discipline of the event*' (2012b: 69). Indeed, this has proven to be the case in varying degrees with all the situations described in this chapter. Thus the operative question here concerns the extent to which gadgets can help in maintaining and building fidelity: how to imagine, construct and maintain a future for the gadget-thing that is true to idea of resonance in line with a commitment to care, becoming and collectivity. In the aftermath of the Egyptian revolution this remains in doubt, given the failure of that revolution and the questions hanging over the social media platforms most associated with it – familiar issues discussed previously around corporate ownership, algorithmic power and exploitation. We know that gadgets can be used to counteract any fidelity to the event, as means of supporting the dominant situation, confusing and obscuring the evental truth. Nevertheless, if we are subjects of events, then in a society saturated by gadgets those events will necessarily be understood and responded to via gadgets as a means of collectivity – and as such as sources of fidelity and subjectivation – for good or ill.

All of the examples I've discussed in this chapter can be understood as forms of gadget action; that is, as actions augmented or shaped in some degree by gadget consciousness. That is certainly not suggest any direct causal relation of gadgets to action, but only to emphasise that collective intention has in different ways been manifested in these actions via gadgets. To what extent the actions explored would have happened anyway, or would have happened differently, is impossible to know.

However, by exploring these examples as frameworks for understanding the character of resonance and idiocy, we have a chance of understanding ways to maximise the former and minimise the latter. As stated at the start of the chapter, this was intended as a heuristic exercise not a scientific or empirical hypothesis. But it is one that provides a framework for projecting forward into the future of gadgets, and in so doing help shape them and us.

6

Gadget Futures

Active social forces work exactly like natural forces: blindly, forcibly, destructively, so long as we do not understand, and reckon with, them. But when once we understand them, when once we grasp their action, their direction, their effects, it depends only upon ourselves to subject them more and more to our own will, and by means of them to reach our own ends. And this holds quite especially of the mighty productive forces of today. (Engels 1978: 38)

In the first decades of the twenty-first century there is a prevalent narrative that the future is not what it once was. There are voices who profess that in fact progress has long been in decline, and that by now we seem to have reached peak misery and live in a state of perpetual collapse and environmental decline, capitalist crisis, austerity and mental destruction. We are witness to the 'slow cancellation of the future', says Franco 'Bifo' Berardi (2011: 18). Elsewhere he wonders if 'there is hope beyond the black hole; if there lies a future beyond the immediate future', and suggests that we live in world where '[t]he sensibility of a generation of children who have learned more words from machines than from their parents appears to be unable to develop solidarity, empathy and autonomy' (2015: 7). This is a view that is in many ways antithetical to the possibilities inherent in the idea of gadget consciousness proposed in this book. While this, one of Bifo's more extreme and glib claims, has very little empirical evidence to support it, beliefs of this kind are widespread and represent something of the zeitgeist, and so need to be addressed. While an important element to any critical theory is a clear-eyed realism about the problems of the present and the path they may take, this should not collapse into the kind of melancholia and despair that is sometimes rendered by such thought. So it is that analysis must also contain hope as we project forward. As such, in this final chapter, and by way of tying up loose ends, I will offer some thoughts about the possible futures of gadgets – this must contain the clear-eyed perspective mentioned, but neither should it be afraid to project a little utopian hope. To do this I will unpack a little further the dystopian strand of thought, not to endorse it but to recognise the seeds of the social concerns it reveals.

GADGET DYSTOPIA

We are constantly surveilled, our desires and tastes are captured and fed back to us in subtly distorted and manipulated forms, we are bombarded by confusing information that disturbs our psyches and undermines our capacity to think and make decisions. Trolls and bots challenge our capacity to connect with others and our labour is shattered into a fractal form that is tracked. Hyper-complex and semi-autonomous algorithms process all this and turn it into big data, both a reification of our subjectivities and an active agent in reshaping and directing us into the mode of dividual. Gadgets are designed to tickle our attention, to probe our brains and prod our amygdalas into releasing pleasure hormones that bring us back to our gadgets like rats in a Skinner Box or willing recipients of digital 'soma'.

Such are the kind of observations which are common in contemporary culture, in particular if you read certain popular commentary on technology by the likes of Evgeny Morozov, Andrew Keen or Susan Greenfield, who all make good headlines proclaiming simplistic disasters, which feed, and feed off, a kind of 'Black Mirror' projection into the future. *Black Mirror*, a TV series written by the critic and commentator Charlie Brooker, presents various frightening and somehow realistic stories based on the development of current technologies. For example, in one episode we are presented with a world in which all aspects of our lives are submitted to an eBay-style rating system wherein social interactions are judged and scored and a running aggregate score for each person is displayed for all to see via their mobile phones. This scenario is extrapolated to a hierarchical social order in which those with lower scores become more and more excluded, with more and more limitations placed upon them.[7] Various other equally dystopian situations are depicted in the series, for example one in which all experiences are recorded via micro body cameras and can be instantly played back to resolve conflicts, supplement forgetfulness, or make

---

7 Such visions are sometimes scarily prescient, as was revealed during the process of writing this book when an eerily similar 'social credit' scheme was unveiled in China.

instant legal judgements. In another episode we see a person's entire social media history and online life being used by an AI programme to recreate a version of that person after their death to comfort loved ones, but in so doing exploiting and manipulating the vulnerability of the surviving family.

What I've intimated is a future projected from the domination of gadget-objects, confronting us as isolated individual subjects, exerting control over us, curtailing our freedoms and turning us into docile and compliant clients. It is easy to see this crystallised in actually existing technology, for example in the Uber app and its kind – installed on a smartphone, the app organises the movements and behaviour of drivers down to the smallest detail, becoming a conduit for rating both customer and driver, for exchanging money, and for recording every detail of the transaction. Yet the Uber app is presented by the company as a force for liberation, sociality and authenticity – the driver becomes his or her own 'boss' and community liaison, thereby realising the capitalist dream of the free contractual relationship as the fundamental mode of all social relations. But this is, as it ever was, pure ideology.

There is clear value in these dystopian narratives as engaging ways to reflect critically on our times and to develop a healthy scepticism about the claims made by companies such as Uber. Nobody creates dystopian narratives as a way to think up future possibilities in order to bring them about. Rather the opposite is true: such narratives are almost always aimed at exaggerating the worst aspects of any society in order to highlight the dangers and try to stop them becoming reality. Thus, there is often a certain similarity between dystopian thought and critical thinking, but like good dystopias effective critique needs to offer some window of hope by pointing us in a different direction – or at least leaving some space for this.

Approaches that just offer blanket cynicism can look like critique – they are often widespread given that they are easy to turn into soundbites and easily digestible binaries – but in fact they are the opposite as they present overwhelming and frighteningly bleak scenarios with little progressive possibility. For example, the clicktivism argument – that gadgets make us think we're doing something, when really we're just submitting to the system – merely encourages us in the conviction that nothing we can do matters, so we may as well sit back and do nothing at all. Again, such an attitude as can be seen in technology writers such as Gladwell or Morozov and Keen, but when examined the substance of these positions is generally just an echo of the cynical knowing tone of the trolls and lurkers on newspaper comments threads.

More helpful critique tries to capture some of the specifics of the situation in terms of radical negativity, but also offers technical and historical insights that contain some seeds for turning the dystopian into the utopian. In that regard the dystopian and the utopian are not opposites or equivalents; rather, the dystopian is a dialectical moment in utopian thought. Here we can introduce a distinction between the critical/dystopian and the cynical.

Scathing arguments made by Byung-Chul Han offer an example of the critical/dystopian. Under the label of 'psychopolitics', Han tells us that 'freedom itself is bringing forth compulsion and constraint' (2017: 1). This is not a negative force of limitation upon freedom, but one in which there is coercion towards positive action, consumption, creativity, innovation and general overachievement. The compulsion to productivity means the subject 'willingly exploits itself without a master'; this is the neoliberal subject as 'the entrepreneur of its own self' (2). The only real freedom, Han tells us, comes from the freedom of the community, of friendship. He makes the point that Marx only ever conceived of freedom as '*self-realisation with others*' and that individual freedom is always a 'ruse – a trick of capital' (3). Yet, according to Han, the current formation of labour makes class struggle impossible because in today's age of isolated '*unlimited self-production*' 'no *political We* is even possible' (6).

Han describes this as 'smart power', which 'cosies up to the psyche rather than disciplining it through coercion or prohibitions' (14); to do this it 'reads and appraises our conscious and unconscious thoughts' (15). The tool that capital uses for this purpose is big data. Going beyond even the Foucauldian idea of biopower exercised over the body, big data invades the psychic realm, providing 'the means for establishing not just an individual but a *collective psychogram*' (21). This would be capital's equivalent of the social endogram, but whereas the latter is the awareness within an individual of the generality of gadget consciousness, the psychogram, in Han's terms, stands for the abstract systemic statistical knowledge of the whole population.

What we can take from Han is an insight into the dystopian potential of a *collective psychogram*, which suggests a situation that if unchecked would look very much like a *Black Mirror* episode in which gadgets have come to outwit and capture us in a totalising and inescapable digital straitjacket. Yet here the critique gives us a diagnosis which leads to a logic of praxis – one facet of the struggle for gadget consciousness must be to maintain the social endogram as a force for recognition and care, to protect it from such dystopic incursions. The imperative is for the integrity of collective intention against the logic of the collective psychogram and the force of 'smart power'; this resistant logic would be represented in relationship with the gadget as thing, and out of that a utopian guiding logic can be born.

Han's approach is really just a development of an idea originally framed by Gilles Deleuze in his 'Postscript on Control Societies' (1995: 177–82). As well as being drawn on by Han, this idea of control is prevalent in much contemporary critical writing on digital technology and culture. Galloway (2004), and Galloway and Thacker (2007), for instance, argue that control resides in the protocols, algorithms and source code that underpin our gadgets and tie together our digital communications systems. Wendy Hui Kyong Chun has also argued that in many respects digital networks follow the logic of control in profound and integrated ways. For example, she tells us that graphical user interfaces (GUIs) encourage acceptance of the logic of neoliberalism among computer users by supplementing the idea of the self-contained rationally driven economic unit; we see this in the way that GUIs help 'move their users from grudging acceptance to feelings of mastery and eagerness' and also help produce '"informed" individuals who can overcome the chaos of global capitalism' (Chun 2011: 8). Chun tells us that 'new media empowers people by informing them of their future' (8). Of course, this is simply a fantasy and far from reality, as she goes on to explain: 'The dream is: the resurgence of the seemingly sovereign individual, the subject driven to know … the dream is the more that an individual knows, the better decisions he or she can make' (8).

So again, the idea is that subjects are captured in a situation that gives an illusion of power and autonomy, but in reality offers the opposite. In her more recent writing Chun has developed this critique to include the analysis of crises as the driving force of new media, arguing that '[c]odes and crises together produce (the illusion of) mythical and mystical sovereign subjects who weld together norm with reality, word with action' (92). Yet even with Chun's analysis, the insight isn't intended to make us throw our arms up in despair, but to alert us to the strategies of power in order to find ways to escape them – in this case the answer being not to fall into patterns that reproduce the logic of bourgeois individualist ideology.

Berardi's writing offers a similar dystopian vein. In *The Uprising* (2012), he critiques the current condition of the 'infosphere' as being 'too dense and too fast for a conscious elaboration of information' (15). The rampant neoliberal deracination of the social has meant that Europe itself has become a 'sad project of destroying, of devastating, of dismantling the general intellect' (39), and that democracy is now under severe threat as 'techno-financial automatisms have taken the place of political decisions' (53). What we see then is a failure of solidarity because cognitive labour has been subjected to 'techno-linguistic automatisms', in a situation in which 'you cannot build solidarity between fragments of time' (55). Again, while this is a rather bleak 'Black Mirror' vision, Berardi still finds an answer to all these concerns in people's capacity to create without subsumption, as hard as that might be. He turns to the idea of thought, language and poetry as a path out of the abyss and recognises the power of the general intellect: 'The new form of life will be the social and instinctual body of the general intellect, the social and instinctual body that the general intellect is deprived of inside the present conditions of financial dictatorship' (157). He further argues for the power of irony over cynicism: 'The cynic wants to be on the side of power, even though he doesn't believe in its righteousness. The ironist simply refuses the game' (166).

Jonathan Beller makes the case that beyond even this what we are now dealing with is an emerging 'fractal fascism interfacing with what has become a kind of platform totalitarianism', and that 'the resources of the senses, the intellect and the will are subsumed and automated in the operations and renderings of "technology itself"', in a process that he calls 'cognitive subsumption by ambient technology' (2018: 5). Beller thinks that this subsumption is written into the DNA of much digital technology and has its roots in the racism and colonialism of the west: 'what we currently call digital culture is actually the second digital culture built atop a first order digitisation by racial capitalism that included colonialism, slavery, hetero-patriarchy and industrialisation' (20).

This control thesis, in which the digital and digital gadgets are captured, is compelling and necessary. It is difficult to deny that we are living in a world of increasing 'fractal fascism', in which the threadbare fabric of democracy is coming apart. However, it is important to note that even for Beller it is the entire ecosystem of capital, plus colonialism, plus digital technology that has led us to this point. As such it is the predominance of this nexus that explains the drift to fractal fascism given that it is 'axiomatic that capitalism and democracy are structurally contradictory' (158). Again, this allows Beller to recognise a route out, a negation of the negation even here. So, for example, he recognises the 'possibility that people no matter their situation could selectively participate in economies of their own choosing' (174).

This is more readily conceivable when we reflect on the gadget brain. If we recognise, as with the 'I' of the strange loop, that one sign system and set of meanings can sit upon another, and to an extent break free to retroactively impact on its own support system, then the gadget assemblage can come to know itself, with all its flaws, and as such develop the will to change itself. This is analogous to, but distinct from, the brain itself, because, unlike the brain, gadget consciousness is able to turn inwards to map, observe and reflect on the substrates of the supporting technical system with which it intersects. This is, again, a deepened aspect of gadget consciousness as a variation of class consciousness and gives it a particularly powerful capacity, especially when that substrate includes the mechanisms of digital capitalism – upon which it sits, but which it can transcend. This enables us to consider alternatives to the grim world in which Uber, Deliveroo, Facebook and Instagram deliver us into the endless unfolding of capitalism as fractal fascism.

GADGET UTOPIAS

It is possible to conceive of a gadget future that is neither the Orwellian vision of a boot stamping down on a human face in perpetuity, nor the discredited optimism and utopianism of the 'Californian Ideology' (Barbrook and Cameron 1995). Something else is imaginable beyond the dystopian dark side – a technology of commonality and community that enables and liberates, that maximises common wealth, experience and freedom. Such a hope has indeed been the source of many alternate visions and experiments. We can trace them back to the imaginings of H. G. Wells and his 'world brain', in which a global repository of all human knowledge would be gathered together and accessed via local terminals available to all, a sure path – Wells thought – to general enlightenment and world peace, noting the imperative that the 'world has to pull its mind together' (Wells 1937).

Jump forward to the early 1970s and another communal hope can be found in Project Cybersyn. This was an attempt to instigate the birth of a socialist technological command and control system for the whole of the Chilean economy, as a way of overcoming the inequalities and inefficiencies of the existing capitalist system. The newly installed socialist president, Salvador Allende, was convinced by Fernando Flores – the technical director of the state development agency, responsible for nationalising Chile's industries – to recruit the British engineer and pioneer of cybernetics Stafford Beer. Flores 'was drawn to Beer's work because of the connection he saw between cybernetics and socialism' (Medina 2014: 32). This could be seen in Beer's modelling of what he called 'The Liberty Machine', 'a sociotechnical system that functioned as a disseminated network, not a hierarchy; it treated information, not authority, as the basis for action, and operated in close to real time to facilitate instant decision making'; thus it 'achieved the balance between centralized control and individual freedom' (33).

When these principles were applied to Cybersyn (a portmanteau word combining cybernetic and synchronise) it meant that '[v]oters, workplaces and the government were to be linked together by a new, interactive national communications network, which would transform their relationship into something profoundly more equal and responsive than before – a sort of socialist internet, decades ahead of its time' (Beckett 2003). Those involved, including Allende,

> saw the system as presenting ways to increase worker participation in factory management. The statistical software evaluated factory performance using a model of production processes. Team members argued that workers should participate in the creation of these models and thus in the design of this technology and in economic management at the national level. (Medina 2014: 6)

As well as integrating production, the interface between the network and the managers was envisioned as a futuristic control room in which the users would sit in high-tech swivel chairs accessing multiple screens and terminals with the two-way flow of information operating as at least analogous with the mode of gathering. As such, Cybersyn indicates the potential for devices purposed with a distinct social intention. Cybersyn was not designed to be fully automatic, to cut humans out of the loop and put algorithms in charge, but rather to generate informational focal points where decisions could be made with real purpose and consequence.

It is impossible here to fully explore the Cybersyn project, but what this quick glance offers is a fragment of an alternative vision of computing.

That vision was most certainly a fork in the road that was not taken in the development of both networked technology and our physical relationship with it. Had it been fully developed, rolled out and socially embedded it could have instilled a very different relation to our devices than the one examined in the previous section.

Nick Dyer-Witheford, in his essay on the possibility of 'Red plenty' (2013), highlights parallel attempts in the Soviet Union to use computers to increase productivity and well-being in relation to the Soviet Gosplan (state planning committee). The Soviets aimed to establish 'a modern computing infrastructure to rapidly carry out the millions of calculations', though the capacity of the computers available made this ambition out of reach, and 'after a decade of experimentation, their attempt collapsed, frustrated by the pitiful state of the Soviet computer industry' (Dyer-Witheford 2013: 4). As with Cybersyn, the Soviet programme was dealing with a system based around large mainframe computers and only with the organisation of industrial production on a national scale. The sheer volume of the calculations needed and the imperative to capture these in a centralised system were both too ambitious and also potentially totalitarian in terms of centralising decision making. When the victory of the 'market' system using decentralised price signals as steering cues was the outcome, the path of neoliberalism was assured.

Yet the development of modern computer power – readily available at marginal cost and widely distributed throughout the population – offers a very different prospect. Dyer-Witheford points out that we do now have the computer capacity to overcome the 'calculability problem', and that you only have to look at the scale of Walmart's operation to appreciate this. Walmart's use of just-in-time ordering and logistical management is a powerful illustration: by the mid 2000s, amongst other massive 'big data' innovations they were already tracking 'over 20-million customer transactions per day' (9).

We are also in a position to offset the dangers of over-centralising power and look to more participatory modes of planning enabled by the massive distribution of devices. Such an economic variation would be a form of socialism more closely associated with the tradition of workers' councils and local assemblies. Socialism could use technologies to plan and orchestrate economies locally, regionally and even more widely. Drawing on Raymond Williams, Dyer-Witheford points out that 'there is nothing intrinsically authoritarian about planning, providing there is always more than one plan' (11). Such an integrated system would thus allow 'a society of participatory, informed, democratic and timely collective planning', and would entail 'fast, varied and interactive communication platforms where proposals could be circulated, responded to, at length or briefly, trends identified, reputations established, revisions and amendments generated, and so on' (12).

Dyer-Witheford doesn't stop there, but outlines an even stronger vision of a potential time of 'plenty' in which the above debates are moot, wherein 'scarcity is replaced with plenitude, ending the need for either prices or planning'. The three tendencies that point towards this are 'automation, copying and peer-to-peer production' (14). Each of these elements would contribute towards the elimination of painful, humiliating and exploitative labour, overcoming the scarcity of goods and the hardships therein, and undoing the alienation of individuals within hierarchical systems of command.

If we take these thought experiments in digital technology backwards – imagining that technology had developed at an equal pace in a parallel universe where Cybersyn had not been discontinued in the wake of the Pinochet coup on 11 September 1973, and where the Gosplan had not been in the hands of a party oligarchy with hopeless technology – then we can readily imagine socialism-enabling gadgets emerging from this process and increasingly strengthening it with careful shared planning and an absence of scarcity. We can picture a world where gadgets are designed not to perpetually distract and reassign attention, to fractalise labour and maximise exploitation, but to augment the fair distribution of work time, to efficiently job share, to provide for the material needs of all and to open up knowledge resources. This would be counter to the current mode of data capture and enclosure behind paywalls and encryption that is designed to extract the tiniest micro-payment from each action and to monetise every glance, gesture and word.

We can imagine in such a world any number of happy gadgets merrily thinging all day long. A socialist Siri or a collectivist Alexa raising the spirit of the common destiny of humankind. Or something like that anyway.

GADGET PREFIGURATION

The question of how we get from here to there is always lurking in the background. The limitations of commercial platforms include restrictions of access to source code, algorithmic management of data, and the conversion of their users' activities into a commodity and the users into providers of free labour, not to mention the global exploitation of the cyber-proletariat (Dyer-Witheford 2015).

Yet capital is still vulnerable because it relies on the revenue generated by users as the core of its business. Commercial platforms have to leave some social interaction that is relatively free and open for their users because they are reliant on them to generate their revenue. The nature of digital capital as parasitic on social labour means it cannot contain or eliminate the processes of communication that fuel and perpetuate the social life. Marx's prediction in the 'Fragment on Machines' that general intellect would be absorbed into constant capital has proved unfounded to the extent that the value-creating power of the human brain has yet to be fully captured by way of a 'real subsumption'. The human brain, with its capacities for invention, empathy and understanding, is therefore an element of the means of production that is deeply elusive to capital. Capital's solution is to instigate a full-spectrum platform biopower. That is, an array of interrelated platforms that attempt to encroach on all aspects of human life, including the general intellect. But capital has failed, even as it has inflicted severe wounds on the brain of labour in that failure. Capital is restricted to a formal, rather than a real, subsumption of the social, so long as elements of social relations remain at least partially inseparable from, and parasitic on, the human brain. These include aspects of unconscious and affective brain activity.

The brain is an absolute limit on the capacity of commercial gadgets to control communication and meaning. Forms of resonance are always possible. It is this absolute limit that provides the antagonistic space for what can be described as a prefigurative zone enabling communicative action and unforced affective flows to take place. Here we can conceive of a more Gramscian tactic in which gadget consciousness must be a vital component.

Thus Facebook, Twitter, iMessage, WhatsApp, Snapchat and a number of other large-scale commercial platforms such as YouTube, Google+, Tubmlr, Digg, eBay, Pinterest – while being fundamentally entrenched in capital economy and functioning towards the valorisation of social labour – still offer opportunities for large-scale connectivity and for deliberation and coordination on a broad scale. Connectivity then provides an opportunity for anti-capitalist political coordination and organisation to take place. There are numerous examples of this, and while I do not intend to revisit the 'Twitter revolution' debate, such zones have been

clearly seen in the use of platforms. This is not to discount the significance, constraints and affordances of matter and code, but to recognise
that platforms are also dependent on the active general intellect that is
gadget consciousness.

Gadget consciousness in this mode can often operate in a cultural frame
to dismantle mechanisms of domination and alienation and reconnect
using a different logic. For example, returning to Berardi, pessimistic as
he otherwise is he makes the point that we can move away from a generalised social aim of 'product growth, profit and accumulation' (2012: 64),
and he has accordingly spoken of our capacity to 'organize a long-lasting
process of dismantling and rewriting the techno-linguistic automatons
enslaving us all' (54). Berardi's specific solution to this problem is a call
to reinvigorate the power of language as dislocated from the exchange-
ability of capital, through a poetic and ironic stance wherein '[p]oetry
is the reopening of the indefinite, the ironic act of exceeding the established meaning of words' (158). In other words, the antagonism through
which the human brain has eluded the real subsumption can be reinvigorated by linguistic forms such as poetry; to put it bluntly: poetry can be
a form of psychic hacking. We can take poetry as only one element of a
much wider array of tools, including digital art and forms of maker and
remix culture. This is, in effect, a practice of countering the effects of
gadget false-consciousness in the cracks that are available now.

One example of an artwork that redirects the subsumption of the
general intellect into gadget consciousness is 'Face-to-Facebook', a work
that was itself based on an exploit. The instigators, Paolo Cirio and Ales-
sandro Ludovico, harvested more than 1 million Facebook profiles using
custom software. Then, using an adapted face-recognition algorithm
they categorised the faces and matched them, much in the way that Mark
Zuckerberg did with his original 'facemesh' algorithm, reworking the
database into a mock dating website. The potential to realise a desired
goal to meet possible partners – the unspoken feature of the platform –
is clearly a move to circumvent the 'safety-first' digital love described by
Badiou. The project's authors tell us that the user's 'smiles will finally reach
what they unconsciously really want: more relationships with unknown
people', but the project also 'starts to dismantle the trust that 500 million
people have put in Facebook'. They also explicitly recognise that 'we are
trying to formulate a simple hack that everybody can potentially use
… that shows, once more, how fragile and potentially manipulable the
online environment actually is' (Ludovico and Cirio, n.d.)

This artwork operates in the mode of hypertrophy, pushing Facebook beyond its limits, re-engineering that which is enmeshed in its desiring circuits. In circumscribing the algorithmic control of encounters it brings to the fore the experience of a local truth that choices are simply a series of forked pathways that undermine the aleatory at every junction. This is a hack, but also a hack in the much more general sense used by McKenzie Wark (2004): a creative act, a moment of generative abstraction that opens a way for new occurrences and things, new connections and ideas. So it is that Face-to-Facebook creates receptivity, preparing the ground for new kinds of subjectivation by providing the experience of usually concealed truths about the experience of online dating and as such priming us for other possibilities.

Another method of prefiguration and of 'thinging' the gadget is a variation on the tactic of exodus. Undertaking digital exodus – that is, the practice of creating, or extracting and repurposing, databases – lies on the borderline between art and illegal action. It is tantamount to creating illegal gadget assemblages, which combine many of the features discussed in this book, and are oriented towards the longer-term nourishing of the social brain and the building of new kinds of utopian imagination and institutions to support that.

AAAAARG.ORG is a publishing platform for the sharing of digitalised books and articles. It is not in the strict sense open, as it is password protected; as such it operates on a tactic of invisibility. However, passwords are distributed on request and the books are offered as a common pool resource to a community that is highly sympathetic towards the principles and the value of open knowledge. It offers a glimpse of both disruption and commoning by its users, taking commonly available hardware, scanners and simple encoding software to turn printed material into PDF format. This then allows sharing of the results, taking advantage of the Web's distributed form and the easy availability of security measures originally designed to protect capital. AAAAARG has also avoided the Web's most centralised control protocol, the domain name system, by simple tricks such as shifting the number of A letters in its URL. By challenging the proprietorial copyright regime the platform is antagonistic, as well as being merely prefigurative. It disrupts through de-commodifying books and making them common, undermining artificial scarcity. This may not appear distinct from the Google Books project as far as its immediate impact on publishing goes, but the longer-term implications are quite distinct as a process of commoning. The accusations and ramifications that have led to AAAAARG being categorised as a 'pirate' operation, and the legal actions against it, are clearly reactions to its threat to the copyright regime.

AAAAARG is also a platform for deliberation, creating a space for discussions on the books it makes common and operating as a platform for the organisation of 'The Public School' – a project for the sharing of knowledge and expertise in a kind of free commons-based university which gives access across the world to texts that would only be available to a select few with access to proprietorial journals and collections in university libraries, with all the attendant privileges and exclusions therein. Of course, it is ad hoc and rather arbitrary, but the collective effect is impressive and offers a great resource for thought; as a template for enhancing the knowledge commons it is compelling. The combination of a tablet computer with a network connection and AAAAARG is very powerful, and transforms the gadget into a very thingly entity. It brings together capacities from across the social and technological spectrum and offers a nexus point of gathering. To return to our Heideggerian language, the fourfold is present within this gadget: it taps into the repository of common wisdom gathered on the AAAAARG servers – which itself represents the shared thoughts and worlds of multitudes of scholars and thinkers – to become a focal point for the attention of the user.

This kind of commoning action matters to capital, as can be ascertained from the very real danger that those undertaking such actions face. The fate of Aaron Swartz – who committed suicide while awaiting trial for hacking into and downloading the archive of academic articles kept on the JSTOR system – is a testament to such dangers. The extreme response of authority, when what is already public knowledge is prised from behind the paywalls of constituted power, reveals the power that such actions have and, again, provides a prefigurative glimpse of alternative worlds.

In the case of both commons-oriented and purely antagonistic gadget assemblages, the question becomes whether they can be maintained and developed without such wide-ranging social changes in time to contribute precisely to bringing these around, given the cost of upkeep in both immediate economic and political terms, in often hostile legal, political and technical contexts. This is exemplified in the push against net neutrality from influential elements within the US government, as well as the recent legal ruling in the UK that has forced a number of ISPs to shut down access to file-sharing websites such as The Pirate Bay and Kick Ass Torrents. There are also a number of smaller ongoing struggles; for example the publisher Verso issued injunctions against AAAAARG, forcing it to remove many titles from its platform. This was somewhat ironic given that Verso has published a number of the recent books exploring the communist hypothesis, including editions by Alain Badiou, Jodi Dean and Slavoj Žižek. Given the massive state, corporate and legal systems at work, the capacity of one individual or group to maintain disruptive or commons-based platforms may not be sustainable – no matter their technical skills. The greater impact of AAAAARG and other such sites may well be in the loss and outrage people feel when they find their assumed right to access and share knowledge – to be part of the general intellect – has been curtailed by legal, state and corporate apparatuses.

In and of itself AAAAARG cannot but help be a deeply ambiguous project, potentially undermining some hard-pressed independent publishers and their revenue models, upon which many writers, editors and publishers rely. What it demonstrates is the need to see such projects not in isolation but as part of a cluster of social and economic factors that need to be developed. One such factor is a universal basic income set at a level that allows for a decent life, or at least mechanisms that allow for the social labour of production to be rewarded in ways that do not simply add to the accumulation of capital and the cycle of exploitation. Ultimately all projects like AAAAARG need to interconnect and become part of a general movement with a concerted vision for change.

This vision should entail: fair distribution of work and leisure time; appropriate access to goods and resources; common pooling of knowledge and education; the challenging of unequal power relations; the elimination of exploitation and the guarantee of mutual respect and recognition of values and identities; an underpinning for the basic dignity of life; collective goals and democratic decision making at all levels; the general opening of opportunities for further flourishing and development. There is a word for all this: communism.

THE IDEA OF GADGET COMMUNISM

While the utopian ideas of technological plenty and freedom should be a marker point on the horizon, communism, at least in its Marxian-inspired strand, also requires a pragmatic and grounded analysis. Thinking about the significance and possibilities for communism in the twenty-first century involves thinking about gadgets as part of a new communist dispositif or ensemble. This is important because gadgets are where the people live, where power lies and where capital is most fully engaged.

The return of communism as a serious political aim was firmly heralded in March 2009, when the conference 'The Idea of Communism' was convened by Slavoj Žižek and Costas Douzinas. Having initially been scheduled to take place in a modest conference room in Birkbeck College, it had to be moved to the Institute of Education's Logan Hall, a 933-seat theatre, which was subsequently supplemented by spill-over video rooms for those unable to secure a place. Even with tickets priced at over a £100 for the three-day event – an irony not lost on a number of the attendees – it was a sell-out. The prominence of the unabashed use of the term communism, and its seeming success, was such that the conference garnered a fair amount of press interest, with *The Guardian* reporting that it was 'the hottest ticket in town' (Campbell 2009).

In the subsequent decade the resurgence of communism has not fulfilled the hyperbole of that event, though we have seen unprecedented gains by the left in some quarters, most notably with the British Labour Party having the first unequivocally socialist leader in its history in Jeremy Corbyn. There has also been the growth of Podemos in Spain and Syriza in Greece, and although neither has managed to achieve its aims, the popular mood has nevertheless been one of increased radicalness and outbreaks of resonant action. Indeed, the idiotic ferocity with which the right and the political establishment have reacted has been a lesson in the threat these shifts in consciousness present.

So it is that the question at the heart of the Idea of Communism conference is still live, as proclaimed by the organisers in the edited collection published the following year: 'whether "communism" is still the name to be used to designate radical emancipatory projects'; the conclusion amongst the conference participants was that 'one should remain faithful to the name "communism"' (Douzinas and Žižek 2010: viii). What this means in practice was judged to be that 'we have to start again and again and beginnings are always the hardest. But it may be that the beginning has already happened, and it is now a question of fidelity to that beginning. This then is the task ahead' (x).

Given the centrality of gadgets to contemporary formations and conceptualisations of identity, self-awareness, social life and activism – as well as the importance of immaterial production to the global economy – the urgency of the debates taking place at conference compel us to extend the question of communism to the heart of our current thinking about technology and what gadget consciousness can lead to. The question concerns the relationship between gadgets as an actually existing realm and the horizon of communist possibility. An engagement with the notion of a communism needs to include a commitment to a direct and concerted political challenge to neoliberalism.

It is in this spirit that I shall explore here what the communist hypothesis offers for the coming of gadget communism. In doing so I will return to the work of Alain Badiou, Slavoj Žižek, Antonio Negri and others. First is the question of whether we must think of gadget communism as the result of a revolutionary event. If so, what role can and should gadgets play in such a revolutionary moment?

To recall, for Badiou the birth of communism hinges on his concept of the event as a rift in the normal fabric of the world that shifts the stable structures of perception, meaning and subjectivity. The event reveals a truth that otherwise would remain covered – the event cannot be predicted, it does not fit into a pre-existing paradigm of understanding, precisely because it exists outside of the prescribed practices and socially and politically legislated modes of existence. The distancing from power thus places communism as a form of anti-power, as escaping the trajectory of the dictatorial variations of twentieth-century communism. A vital aspect of this is the mode of subjectivity that is evoked in this kind of thought, a kind of subjectivity committed to collective action and solidarity in a common cause:

> The communist hypothesis is that a different collective organization is practicable, one that will eliminate the inequality of wealth and even the division of labour. The private appropriation of massive fortunes and their transmission by inheritance will disappear. The existence of a coercive state, separate from civil society, will no longer appear a necessity: a long process of reorganization based on a free association of producers will see it withering away. (Badiou 2008: 34)

While I have discussed the potential of the event in the context of gadgets in relation to uprisings, such as the Egyptian revolution in 2011, this exploration was limited to the idea of gadget action – what is set out above is more all-encompassing. The question now is whether a longer-term social transformation, true to the communist hypothesis, is possible, sustainable or indeed desirable.

In line with this aim Badiou proclaims a set of invariants that reflect deeper communist currents manifested in different ways at different times: 'intellectual patterns, always actualized in a different fashion' (35). They are, as Jodi Dean captures them, 'the egalitarian passion, the Idea of justice, the will to end the compromises with the service of goods, the eradication of egoism, the intolerance towards oppression, the desire for the cessation of the State' (2012: 180). Bruno Bosteels tells us that they have their roots in 'the universal aspiration of the exploited to topple every principle of exploitation and oppression' (2011: 277).

For Badiou, while the communist hypothesis is under sustained attack and denial by established power, especially in this period of disorientation, it can still be 'defined by three axioms' (2018: 3). Firstly, 'the egalitarian idea', which has come under pressure from the ideology that human nature 'dooms us to inequality'; the latter must be overcome with the recognition that collective action must be 'consistent with the communist hypothesis' (3). The second axiom is that the 'existence of a separate, coercive state is unnecessary'; this is akin to the idea of the withering of the state, which entails that 'we organise popular political action without subjecting it to the idea of power', that is, of 'representation in the state' (3). The third axiom states that the division of labour, which supports classes and hierarchies, be overcome; thus 'we should – and we can – aim at an essential polymorphy of human labour' (4). What these hypotheses represent is 'not a programme but *maxims for orientation*' (4). Badiou believes that we are 'at the very beginning' of a new sequence (5). To pursue this we need to hold on to an 'impossible' point that is not present in the current situation but is nevertheless a 'real point'.

The proposition of gadget communism may seem ludicrous within the current formulation of wealth and power, yet evental thinking offers a shard of hope that things can, and will, change. However, given the proprietorial and private nature of most platforms and gadgets within already existing capitalist systems, this is profoundly difficult. But I will argue that it is not impossible – even if this means proposing a slightly weaker conception of the event.

The event requires something radically new to enter the world, something unknown and unknowable. Control thus presents a fundamental problem for a Badiouan politics of digital rebellion, given the parameters of digital events that, like any other, necessitate the radically new. If the digital realm is fundamentally characterised by its prescriptive nature then the realm of the digital has become defined, to use Richard Grusin's term, as one of 'pre-mediation'. Grusin (2010) argues that media, and digital media in particular, now truncate or short-circuit the possibility of events entirely. Consumers of media are framed within a set of technical and semiotic boundaries that keep them within the scope of acceptable possibilities, of choices within the prevailing political parameters of not only actions but also affects. If all possible pathways are being chased down by processes of premediation, then decisions are based either on a movement along algorithmic pathways, whose parameters are by definition already pre-empted, or on affective responses that have become embedded in unanswerable preconscious iterations of cybernetic self-comforting.

With premediation we can see not only digital networks in their own terms, but also the extent to which they have become entwined with a wider military-industrial-entertainment complex. Badiou himself, in his exegesis on love, hints at the difficulty of a platform event when he discusses the process of online dating. According to Badiou, dating sites offer only an antiseptic version of love, a 'love comprehensively insured against all risks' (2012a: 6). He associates such a love with the promise of a '"zero deaths" war' (8), wherein the risks are all systemically offset and the daters 'won't find it difficult to dispatch the other person if they do not suit' (9). While it is not overtly stated, the conjecture is that a dating platform filters out all contingencies and possibilities for encounter. While such filters obtain in all kinds of situations, in a protocological digital network the algorithm that controls selection processes and eliminates those unsuitable from view institutes a material bar from the exposure to chance. 'Safety-First love, like everything governed by the norm of safety, implies the absence of risks for people who have a good insurance policy' (9). The same logic is in operation across all major social networking platforms, as they maintain strict protocological limits on encounters and gather the processes of linking and distribution under a single prescriptive proprietorial framework and patterning. Yet it is precisely here, in the singular framework of protocol, that openings to subvert this risk-adverse logic are always present.

The possibility that a purely digital or gadget-centred event might happen is therefore highly limited, but it increases to the extent that we consider the whole assemblage, accounting for all the interlinked brains and the collective will of gadget consciousness. In what sense can we understand a gadget as offering the chance for a rupture of the presented world and the breaking through of a truth? The control situation described above can be challenged if the dispositif itself can be hacked and re-directed from control to freedom. We also need to ascertain whether capital's constant crises and systemic anomalies always produce control, or whether cracks and spaces can nevertheless appear – whether from within code itself or from elsewhere – and as such if platforms open up possibilities for the radically new. Beyond this we need to ask if an avenue, or perhaps even a line of flight, can open up onto the communist horizon.

One model that offers such a potential is the use of the 'exploit'; that is, a systemic flaw, break or even opening that can be worked at, pushed and leveraged against the system itself. The concept is taken up by Galloway and Thacker, in the wake of the innovative practice of hackers. They argue that 'within protocological networks, political acts generally happen not by shifting power from one place to another but by exploiting power differentials already existing in the system' (2007: 81). These exploits include the power of viruses or worms that often do not damage systems but rather find paths and ways to use the protocological controls against themselves. Often such exploits also generate emergent effects, evolving from within systems and acting as non-human agents. While recognising that entities such as viruses and worms are not a concrete model for 'progressive' politics, Galloway and Thacker argue that they do give us a glimpse of both 'the plasticity and fragility of control in networks' (95).

One key tactic for resistance that Galloway and Thacker induce from this logic is that of disappearance: to become hidden in the society of control is to short-circuit its capacity to accumulate data. Seb Franklin builds on Galloway and Thacker's theory of the exploit, telling us that it is exactly in the ambiguity of being unclassifiable as either user (consumer) or producer (labourer) that resistance can be found. Thus it is 'not a question of hiding, or living off the grid, but living on the grid, in potentially full informatic view, but in a way that makes one's technical specification or classification impossible' (Franklin 2009). This might include the simple flashing of an infra-red beam into a camera, or the practice of 'circuit bending' in which technologies are diverted and misdirected, not with highly technical programming but through rudimentary hacking using only basic technical knowledge. It would also include practices of becoming anonymous, of encryption and the use of 'dark' nets based on software such as Tor. These are consistent with Galloway and Thacker's taxonomy of the exploit as passing through the stages of Vector, Flaw and Transgression (2007: 97). Finn Brunton and Helen Nissenbaum offer a full taxonomy of tactics for 'obfuscation' (2015), and indeed there is a whole noble tradition of hacking and hacktivism (Jordan 2008, 2016) that has evolved into movements for change including that of Anonymous. What these examples indicate is that collective acts of intention can push a structure towards a stress point that can provoke, if not an event, then a shock from which opportunities for the creative construction of alternatives can develop.

Yet disturbance and shocks to the networks can also come from outside or emerge from unexpected and unplanned directions. The refugee crises taking place in the Mediterranean since the Syrian war has been a stain on the European Union, with over 2,000 migrant deaths at sea in the first half of 2017 alone (Deardon 2017). This has been a largely invisible scene of suffering, with the western press focusing on the migrants as sub-human invaders. They are precisely what Badiou refers to as the excluded part, a group who are of the situation but are not visible in it. On 2 September 2015, Alan Kurdi, a three-year-old boy, drowned at sea as his family crossed the Mediterranean from Syria to the Greek island of Kos. His body, dressed in a red T-Shirt and shorts, was photographed lying face down on the beach. His appearance was similar to any European child of his age, and the powerful and horrifying image flashed around the world. Helena Smith reported that 'Within hours it had gone viral becoming the top trending picture on Twitter under the hashtag #KiyiyaVuranInsanlik (humanity washed ashore)' (Smith 2015). The image then went on to dominate TV news coverage and covered the front pages of the world's newspapers the next day. The UK's *Daily Mail*, notoriously right-wing and racist, had a full-page cover photo of the child being picked up by a police officer, with the headline, 'Tiny victim of a human catastrophe'. Just a few months earlier, on 31 July, the *Mail* had carried a front page with images of migrants climbing onto lorries in Calais with the headline 'The "swarm" on our streets' – one of many such front pages and typical of the UK's dominant right-wing press.

For a period after the tragedy the entire media and government discourse around migrants shifted from one of hostility and dehumanisation to one of recognition and care, in which the talk was of an awakening, of everything having changed. The appearance of the image of Alan Kurdi had forced the invisible excluded part into the situation, no longer present merely as a blurry mass of bodies in a truck or standing by a fenced-off Eurostar terminal. This was done via the ecology of gadgets: snapped on a smartphone, the image was distributed via social media and remediated across the spectrum of media platforms into the networks of debate, affect and response.

However, what this and the other crises discussed demonstrate is that, without fidelity, without reacting mindfully and carefully and collect-edly, such moments can come to nothing. While there was a tangible and immediate shift in the discourse, this has not lasted and has not been translated into policy; indeed, anti-migrant rhetoric and the suffering of migrants has increased. Just a year later, Patrick Kinglsey reported that 'Europe has gradually abandoned the humanitarian approach of last winter' (Kingsley 2016). Two years on at least a further 8,500 refuges have died crossing the Mediterranean Sea (Dehghan 2017). The moment in 2015 represented a forking between the gadget-thing and gadget-object orientations. The shock of the image presented the chance to force through profound changes, following the entry of a previously excluded group into the situation and the responsibilities and commitment that would entail. Yet this opening was forcibly closed again as the 'reality' re-imposed itself. In that regard the 'event' did not suffice.

Shocks, disappearances and exploits ultimately still operate within the perimeters of the control systems of protocol, while images of suffering and war are eventually absorbed into the spectacle and lose their potency; as such they are more akin to a temporary internal tremor or remodulation. Such hacks and exploits may well intervene in the operation of control long enough to create disruptions that escape from behind the surface of presented reality, but may themselves not be events. They are rather shocks that bend, stretch and rupture, characterised by their unpredict-able ramifications and knock-on effects, including unforeseen emergent features. Any protocological network 'event', in that sense, may not be a 'pure event' as such, but is just such a shock – a spanner in the smooth systemic modulations of probability that gets pushed to an extreme to ripple out across the entire techno-capitalist dispositif. Without already existing organisations, solidarity structures or the collective will to act on a grand and decisive scale, the chance of communism coming to fruition though technology alone is non-existent.

Nevertheless, what such practices do indicate is the possibility of openings in the control protocols, in the mode of operation of gadgets as objects. Those who render themselves invisible or unclassifiable, or who force themselves into the horizon of recognition, all exert will and agency to contribute to such a politics. To put it another way, disappearance and the hidden actions undertaken behind the evental horizon are a form of resistant premediation, contributing to the forcing of what Galloway and Thacker would call impulsion or hypertrophy. This practice is designed not to resist technology but to push it 'into a hypertrophic state, further than it is meant to go' (2007: 98).

This shares a certain sensibility with the idea of 'accelerationism'. The key advocates of this view, Nick Srnicek and Alex Williams, argue that an accelerationist hypothesis 'takes an existing capitalist tendency and seeks to push it beyond the acceptable parameters of capitalist social relations' (2015: 109). For them this entails a shift towards a new kind of society that would include reductions in the working week and a variation on the Red Plenty thesis of wealth for all. They claim that this view is a development of and replacement for what they call 'folk politics', which is characterised by small actions with 'resonances' and which they suggest means 'the strategic imperatives to expand, extend and universalise are left unfulfilled' (35). However, when they come to offer their own thesis as to how the shift from micro to macro is to take place we are left with the notion that pushing existing technology to a limit that will somehow miraculously bring about this change; all they are able to suggest beyond this is the construction of a new prefigurative version of hegemony. This then results in a rather Trotskyist position where we would just spend our time doing 'preparatory work for moments when full-scale struggle erupts' (132).

The problem is that Srnicek and Williams' understanding of technology is based on advocating automation predicated on the current logic of capital, in which efficiency is the primary directive. This implicitly replicates Albert Borgmann's view of technology as an external entity that can and should only support an otherwise organic life of plenty. By dis-locating intention from the technological ensemble itself, they inevitably don't differentiate between technological things and objects. While I agree that local and micro-level action needs to include continuing fidelity and institutional support on a bigger scale, this can and should draw on and develop the resonances that hold together collectives on a horizontal plane. To dismiss all such collective action as 'folk politics' seems highly counterproductive – in the end what they advocate sounds very similar anyway, in the form of advocating for a 'diverse ecosystem of organisations' (162). To suggest a gadget communism is to recognise the importance of hypertrophy or acceleration as a tactic, but the need to go further than this is imperative as a strategy.

There is another strand of contemporary communist thought that can be helpful here, recognising the importance of evental thought, but offering a slightly lower bar. This is identified with the political philosopher Slavoj Žižek – influenced by Lacanian psychoanalytical thinking and elements of Leninist political philosophy – and can helpfully further supplement the idea of gadget communism. For Žižek the working through of an intrinsic antagonism would be the necessary element – the act of pushing the antagonism to breaking point, of forcing it beyond an internal contradiction into a generalised revolutionary situation, but recognising that it must also lead to a further synthesis. In this sense Žižek places himself firmly within the tradition of dialectical materialism.

Žižek identifies four profound antagonisms in contemporary capitalism: ecological catastrophe; private property as the predominant form of intellectual property; new techno-scientific developments; and new forms of apartheid (2009: 91). His distinctly dialectical position entails breaking open these antagonisms and forging them into class positions capable of creating the communist moment. We can detect these antagonisms present in gadgets in a number of ways too: in the tension between gadgets as things and objects; in the tension between gadgets as objects and nature; in the tension between public and private in gadgets as platforms.

Here we can draw on Žižek's interpretation of a Leninist commitment, in that he sees the multiple coagulations of elements gathering around the fourth antagonism, which includes the exploitation of labour. It is this antagonism that capital cannot do without, even when the others may be overcome in variations of socialism and communitarianism. We can see this in gadgets also, in all of the five uses of gadgets by capital identified in Chapter 2. Herein lies the specific need for communism and the need to be wary of overly reformist solutions. One example is any attempt at the resolution of an antagonism within the frame of liberal hegemony and the parliamentarianism of neoliberal democracies. This is a hegemony that proclaims its constant support for freedom, but in reality is one of the greatest mechanisms for presenting a 'formal freedom', in the sense that Lenin used the term, while proscribing any kind of 'true choice'. This is so because 'Formal freedom is the freedom of choice within the coordinates of the existing power relations, while actual freedom designates the site of an intervention that undermines these very coordinates' (Žižek 2002a: 544).

Thus, in the current climate it is tempting to step back from action given that it 'will be an act within the hegemonic ideological coordinates' (545) and as such remain within a 'certain limit'. Rather, according to Žižek, 'to reinvent Lenin's legacy today is to reinvent the politics of truth' (547). What this means for Žižek is not an abstract truth of transcendent knowledge, or one of negotiated compromise, but precisely that of complete one-sided commitment. This is in contradistinction to the proclaimed range of current tendencies in left thought. The abiding tendency is that of the comfortable intellectual indulging in the 'narcissism of the lost cause', in deconstructive thinking in which the moment of realisation of communism remains forever deferred as a 'dream of presence' (2009: 88). Rather, what we see in practice is that 'all successful revolutions ... followed the same model, seizing a local opportunity in an extreme and critical situation' (89). In other words, this is the radical application of intention.

Žižek sees in Lenin the capacity to shock, to act with faith on a revolutionary path even when the prevailing conditions are against this, even if the party begs to differ. Thus, Lenin stands for 'the compelling freedom to suspend the stale, existing (post)ideological coordinates' (2002c: 554). However, this is not to revive the great man of history thesis or to fetishise the vanguard party. In the introduction to his selection of Lenin's writings, Žižek argues that while bypassing the intransigent party Lenin tapped into a 'revolutionary micropolitics' which instigated 'the incredible explosion of grass-roots democracy, of local committees sprouting up all around Russia's big cities ... taking matters into their own hands' (2002b: 7).

The key is activating the moment of shared vision, instigated as the world undergoes a major rupture. This is something of an inverse variation on Naomi Klein's Shock Doctrine thesis, in which the 'taking matters into their own hands' becomes the imperative (Klein 2007). This is not, Žižek argues, a utopia for a distant moment but 'the urge of the moment is the true utopia', and in that moment the imperative is to 'invent a new communal social form without a standing army, police or bureaucracy, in which all could take part in the administration of social matters' (5). For Žižek, Lenin's greatness lies in his forging of the moment for revolution; in the wake of the disaster of 1914 and against majority opinion, 'he wasn't afraid to succeed', so that 'instead of waiting until the time was ripe, Lenin organized a pre-emptive strike' (6). Ultimately Žižek's reading of Lenin places truth as a form of political fidelity and communism as a political act of rupture, a breaking free of the very conditions of constraint, thinking beyond the edge of the actually existing politico-economic universe.

If we return again to the work of Michael Hardt and Antonio Negri and their 'joy of being communist' (2000: 413) then here also the place of antagonism is key. In his contribution to the 'Idea of Communism' conference, Hardt describes the increasing hegemony of immaterial labour and production and sees its development as one that 'returns to centre stage the conflict between the common and property as such' (2010: 135). Working through the contradictions of cognitive capitalism now means that capital no longer creates value through profit but through a return to rent: 'patents and copyrights, for example, generate rent in the sense that they guarantee an income based on the ownership of material or immaterial property', the key point being that 'capital remains generally external to the processes of the production of the common' (137). The use of rent is a way of valorising the common, without capital intervening in the production process and undermining its productivity. It provides the conditions for the multitude to extract itself, and yet at the same time explains the increasing securitisation of the state: ever more modes of control, ever more draconian forms of policing and repressive violence. Finance, Hardt tells us, 'expropriates the common and exerts control at a distance' (138).

Given this development, the creation of a gadget communism must ultimately rest on throwing off the capacity of capital to extract rent, and this also includes cognitive rent. The first difficulty with regard to this is that the forms of rent are not always obvious, as the mechanisms of valorisation are profoundly enfolded in everyday social life. Yet capital still contains the seeds of its own destruction – not automatically, but 'through the increasing centrality of the common in capitalist production – the production of ideas, affects, social relations and forms of life – are emerging the conditions and weapons for a communist project' (143). This is essentially what I described in terms of the various forms of resonant action and the power of gadget consciousness therein. This is resonance not as 'folk politics' but as a turn towards solidarity and antagonism that finds its home in becoming common.

Indeed, Hardt is keen to reiterate the importance of retaining the word communism as part of this struggle, so as to resist the reduction of the idea to the definition given by its opponents; he tells us that it is 'important for us to recognise alternatives within the tradition and affirm the streams we value most. We thus feel the need to struggle over the concept of communism and insist on what we consider its proper meaning' (2012). To contribute to the realisation of a truly 'full' communism, gadget consciousness needs to become part of a greater revolutionary process in which the power to valorise and absorb creativity is wrestled from capital. In their summary of the key themes from the Idea of Communism conference Douzinas and Žižek argue that, above all,

> Neoliberal capitalist exploitation and domination takes the form of new enclosures of the commons (language and communication, intellectual property, genetic material, natural resources and forms of governance). Communism, by returning to the concept of the 'common', confronts capitalist privatizations with a view to building a new commonwealth. (2010: xi)

This commonwealth should aim to 'bring about freedom and equality. Freedom cannot flourish without equality and equality does not exist without freedom' (x). This is in line with Badiou's communist invariants; the commitment to the key invariant, the ending of exploitation and oppression, is an element of an ongoing historical movement in which the digital age must be included and which gadget consciousness must aspire to supplement, enhance and enable. As Žižek argues, '[w]ithout the World Wide Web socialism would be impossible ... [o]ur task is here merely to lop off what capitalistically mutilates this excellent apparatus, to make it even bigger, even more democratic, even more comprehensive' (2002b: 17).

While there are a number of differences, the similarities between the Badiou, Žižek and autonomist variations of communism need to be noted. These are significant for thinking through the multifarious possibilities of gadget communism. Žižek does not offer a positive prescription of the shape of future communism, which would undermine his fundamental commitment to communism as processual, in line with 'Marx's notion of communism not as an ideal, but as a movement' (2009: 88).

But what we see in Žižek, and indeed in autonomism, is a remnant of the subject as collective agent re-emerging. For example, when Žižek argues that the antagonisms also have in common 'the process of proletarianization, of the reduction of human agents to pure subjects deprived of their substance' (99), this implies the existence of such a substance, or rather quality, and the power of its return. The reversal of this exclusion activates the 'part of no-part', in which the excluded return to represent the universal. We have, Žižek tells us, 'a name for the intrusion of the Excluded into the socio-political space: democracy' (99). This also chimes with Hardt's acknowledgement of the fundamental antagonism between the common and rent, with the former being subsumed and obscured by the latter in the current capitalist configuration. It is the action of the multitude that is needed to overcome this contradiction, to reclaim the common. Such a return of the excluded is in line with and supported by the agency of resonant collectives in gadget consciousness.

The agency of the collective offers a window to reimagine the place of the party. It is difficult to imagine a consolidation of and fidelity to the process of becoming common without some supportive political form to gather and sustain it. In her book *Crowds and Party* Dean offers an argument for how to move beyond the impasse of recent struggles, for example that of the Occupy movement, which she says was captured by the individualism and fragmentation of communicative capitalism: 'individualism ... undermined the collective power the movement was building' (2016: 4). I agree with this diagnosis, which reflects the discussion on idiocy in the previous chapter, apart from the already discussed dismissal of the role of digital networked communications. Dean argues that the party gives shape and endurance to the politics of resistance and offers an alternative mode of collective subjectivation to the individual. Once changed, once ignited, the party – Dean opines – creates the political force, the antagonism, to sustain change. This is against the attempts to subsume this impulse into the logic of communicative capitalism, as expressed in concepts such as the 'smart mob' or the 'wisdom of crowds', which deny the political potential of the crowd by reducing it to a social salve. By integrating individuals into a digital meshwork of aggregated productive units – but without the torsion that Dean describes as binding individuals into a collective (as described in Chapter 2) – nothing will happen. Such is the 'technophilic imaginary' (15) of liberation within capital.

There is an important truth here, but again Dean's implicit critique of digital communication as such creates problems for thinking about the party form in this context. However, I suggest the possibility of gadget consciousness actually supports the possible reformulation of the party.

Gadget consciousness is able to support the torsions of the crowd into the party precisely through recursion and the creation of long-term resonances that flow through and in parties to reshape and open them to becoming.[8] In that sense the torsion of gadget class consciousness would be precisely this mechanism towards becoming the party and the party as becoming – a mechanism for embedding care, becoming and collectivity.

YOU ARE A GADGET

Gadget communism as a force for care, becoming and collectivity can be imagined – indeed we have already come some way to thinking it. Badiou's three communist axioms of equality, the withering of the state and the overcoming of the division of labour, also underpin this. Badiou's first maxim is in line with the loopiness and resonance of collectives which operate in recognition of the equality of actors. This is conceivable precisely in the scenario in which gadget consciousness augments and articulates with collective action consistent with care as a form of recognition, and which works with the orientation of gadgets as things that gather. It is precisely a consciousness of the social endogram that steps away from a purely individuated response that I characterised as the idiotic hypothesis; indeed one can posit the idiotic variation as the inverse of the communist hypothesis. We can imagine the breaking down of proprietary systems that target users as commodity units so they are instead used to distribute and enhance access to, for example, medical services, education, democratic deliberation, goods and services, and the learning of skills in the development of gadgets themselves.

Secondly, the withering away of the state – at least in terms of its command or authoritarian aspects – can also be readily conceived through the lens of the gadget committed to becoming. Here experi-

---

8   There is a lot more that could be said here, but a full-scale reappraisal of the party form is beyond the scope of this book. However, by way of example, we can see evolving chains of dialogue and recognition forming into clusters of networked coordinated actors in the Momentum movement within the UK Labour Party – a process that could be considered a developing form of gadget consciousness, and that has operated to challenge and contain some of the more hierarchical and ridged aspects of the party form. Of course, this has created a degree of contestation and resistance in the Labour Party and itself contains a number of contradictory elements and tendencies. But perhaps this itself can be seen as resonance and polyphony in action towards coordinated goals and an overcoming of dialectical tensions. This is undoubtedly an ongoing issue and one for further debate beyond these pages.

ments such as the X-net platform in Spain or the kind of 'Red Plenty' scenarios discussed above can operate as templates for imagining a mutually organised and cooperative mode of interaction and social organisation that enhances and develops all who participate. The state's functions can be distributed into collective decisions and designed down to local levels the better to meet local needs, while also being coordinated with national and international systems of production and distribution to maximise collective benefit – promoting maximum efficiency not in terms of capital but in terms of promoting other democratically arrived at goals such as fairness and environmental protection. This is a notion of a digital commonwealth that is orchestrated through gadgets and is responsive to gadget consciousness. It entails a reinvigorated and reimagined mode of democracy.

The process of the withering away of the state also overlaps with the third and final maxim of unpicking the division of labour. Here the affordances of gadgets can be profoundly helpful. I discussed in previous chapters the idea of gadget class consciousness as something that develops Marx's notion of the general intellect. Instead of the knowledge and skills of the workers being frozen into the machines of production, we see an active general intellect as a force of liberation to redirect them. Where gadgets are themselves the means of production the active general intellect can requisition them into the service of the collective good. We can imagine this emerging directly out of the current evolution of capitalism, much in the way that Marx discusses the industrial era giving birth to its own downfall. So, for example, Uber becomes a common platform to coordinate transport, integrated with public transport systems rather than competing with and undermining them, thus working to communise what was individuated and eventually to decompose the system of individuated atomised transport. Amazon becomes a way of distributing and managing resources for the satisfaction of basic needs, and with the increasing efficacy of production and distribution becomes perfectly capable of distributing luxury goods equitably. Many goods and services could, and should, be distributed as common goods rather than commodities: foods, medicines, clothes, care and support services, and so on. Apple becomes an organisation dedicated to innovating new ways to ever more successfully turn gadgets into focal things. The imperative of such reimagined organisations, arrived at through the common will of gadget consciousness, would be to increasingly leverage and ratchet up the benefits of that common will.

Leveraging the common will for the common benefit of all would be a suitably recursive goal, but with a genuine application of Steve Jobs' original marketing line from the famous 1984 advertisement for the Macintosh computer: it would be a revolution. These capitalist gadget-oriented organisations have become ever more dominant and have a single point in common: the artificial maintenance of the division between consumers and producers and of the division of labour. Artificial scarcity is a product of the need to generate surplus value, and this is even more true of digital goods that can be replicated at near zero cost. The elimination of these divisions would be a huge step to fulfilling the aims of the communist hypothesis.

What remains vital for a broader communist hypothesis is the resonance of action, the pragmatics of coordination and the movement beyond prefigurative zones towards a large-scale commons. Thus, the aim of a 'full' gadget consciousness must include, but also go beyond, the disruption and hypertrophy of existing commercial gadgets towards gadgets of common organisation, production and distribution of the surplus for the common good. The facility for the broader infrastructure to support the will of gadget consciousness needs to be established, expanded and materially maintained – while being permanently dislocated from capital. Whatever the nature of any specific gadgets to come, we can be sure it is in becoming common that we will find the most powerful opening for realising the power of gadget consciousness.

These are not technically difficult problems; the obstacles are only political and economic. Therefore, what is indisputably necessary for an ongoing gadget communism is, firstly, to act and to communicate, to move towards subtraction from capital; and, secondly, to produce 'common' platforms capable of sustaining the communist horizon as a living reality – to build spaces, places, subjectivities (in Badiou's sense) which generate the momentum for exodus and for subtraction from capital, and in the long-term help in healing the psychic wounds capital inflicts. Such a gadget ensemble would constitute a counter-dispositif. Vitally, this counter-dispositif needs to incorporate the world as a whole – gadgets alone will do nothing.

Bibliography

Agamben, G. (2009) *What is an Apparatus and Other Essays*. Stanford: Stanford University Press.

AP News (2014) 'Spain's "Xnet" Corruption Fighters Expose Graft', 14 December, https://apnews.com/a67563e720d84d86943974e6e7befac2.

Baars, B. (1997) *In The Theatre of Consciousness*. Oxford: Oxford University Press.

Badiou, A. (2005) *Metapolitics*. London: Verso.

Badiou, A. (2006) 'The Saturated Generic Identity of the Working Class', chtodelat.org: http://tiny.cc/bq83fx.

Badiou, A. (2008) 'The Communist Hypothesis', *New Left Review* 49: 29–42.

Badiou, A. (2009) 'Thinking the Event', in P. Engelmann, A. Badiou and S. Žižek, *Philosophy in the Present*. Cambridge: Polity.

Badiou, A. (2010) 'The Idea of Communism', in C. Douzinas and S. Žižek, *The Idea of Communism*. London: Verso.

Badiou, A. (2012a) *In Praise of Love*. London: Serpents Tail.

Badiou, A. (2012b) *The Rebirth of History*. London: Verso.

Badiou, A. (2018) *Greece and the Reinvention of Politics*. London: Verso.

Ball, J. and Brown, S. (2011). 'Why Blackberry Messenger was Rioters' Communication Method of Choice', *Guardian*, 7 December.

Barbrook, R. and Cameron, A. (1995) 'The Californian Ideology', *MetaMute*, 1 September, www.metamute.org/editorial/articles/californian-ideology.

Barker, G. (dir.) (2015) *The Thread* [film].

Baudrillard, J. (2010) *The Agony of Power*. New York: Semiotext(e).

BBC News (2011a) 'England Riots: Twitter and Facebook Users Plan Clean-up', 9 August, www.bbc.co.uk/news/uk-england-london-14456857.

BBC News (2011b) 'London Mayor Boris Johnson Carries Riot Clean-up Brush', 9 August, www.bbc.co.uk/news/uk-14462001.

Beaumont, P. (2011) 'The Truth About Twitter, Facebook and the Uprisings in the Arab World', *Guardian*, 25 February, www.theguardian.com/world/2011/feb/25/twitter-facebook-uprisings-arab-libya.

Beckett, A. (2003) 'Santiago Dreaming', *Guardian*, 8 September, www.theguardian.com/g2/story/0,3604,1037327,00.html.

Beller, J. (2018) *The Message is Murder: Substrates of Computational Capital*. London: Pluto Press.

Berardi, F. (2009) *The Soul At Work*. Los Angeles: Semiotext(e).

Berardi, F. (2011) *After the Future*. Edinburgh: AK Press.

Berardi, F. (2012) *The Uprising: On Poetry and Finanace*. Los Angeles: Semiotext(e).

Berardi, F. (2014) *Neuro-Totalitarianism in Technomaya, Goog-Colonization of Experience and Neuro-Plastic Alternative*. New York: Semiotext(e).

Berardi, F. (2015) *Heroes, Mass Murder and Suicide*. London: Verso.

Beyer, J. L. (2014) *Expect Us: Online Communities and Politcal Mobilization*. Oxford: Oxford University Press.

Bland, A. (2014) 'Firechat: The Messaging App that's Powering Hong Kong Protests', *Guardian*, 29 September, www.theguardian.com/world/2014/sep/29/firechat-messaging-app-powering-hong-kong-protests.

Bor, D., Dijk, D., Foster, J., Gottfredson, L., Jackson, A. and Mather, G. (2017) *How Your Brain Works*. London: John Murray Learning.

Borgmann, A. (1984) *Technology and the Character of Contemporary Life*. Chicago: University of Chicago Press.

Bosteels, B. (2011) *Badiou and Politics*. Durham, NC: Duke University Press.

Bratton, B. H. (2015) *The Stack: On Software and Sovereignty*. Cambridge, MA: MIT Press.

Brunton, F. and Nissenbaum, H. (2015) *Obfuscation: A User's Guide for Privacy and Protest*. Cambridge, MA: MIT Press.

Caffentzis, G. (2013) *In Letters of Blood and Fire*. New York: PM Press.

Campbell, D. (2009) 'Move Over Jacko, Idea of Communism is Hottest Ticket in Town this Weekend', *Guardian*, 12 March, www.guardian.co.uk/uk/2009/mar/12/philosophy.

Castella, T. (2011) 'England Riots: Are Brooms the Symbol of Resistance?', *BBC News Magazine*, 10 August, www.bbc.co.uk/news/magazine-1445741.

Castells, M. (1997) *The Rise of the Network Society*. Oxford: Blackwell.

Castells, M. (2009) *Communication Power*. Oxford: Oxford University Press.

Caudwell, C. (1977) *The Concept of Freedom*. London: Lawrence & Wishart.

Chalmers, D. J. (n.d.) 'Facing Up to the Problem Of Consciousness', http://cogprints.org/316/1/consciousness.html.

Chun, W. (2011) *Programmed Visions*. Cambridge, MA: MIT Press.

Coleman, G. (2015) *Hacker, Hoaxer, Whistleblower, Spy*. London: Verso.

Costandi, M. (2016) *Neuroplasticity*. Cambridge, MA: MIT Press.

Crogan, P. (2010) 'Knowlege, Care, and Transindividuation: An Interview with Bernard Stiegler', *Cultural Politics* 6(2): 157–70.

Cunningham, G. W. (1908) 'The Significance of the Hegelian Conception of Absolute Knowledge', *The Philosophical Review* 17(6): 619–42.

Curtis, N. (2013) *Idiotism, Capitalism and the Privitisation of Life*. London: Pluto.

Damasio, A. (2000) *The Feeling of What Happens: Body, Emotion and the Making of Consciousness*. London: Vintage.

Damasio, A. (2006) *Descartes' Error: Emotion, Reason and the Human Brain*. London: Vintage Books.

Damasio, A. (2010) *Self Comes to Mind*. London: Heinemann.

Dean, J. (2009) *Democracy and Other Neoliberal Fantasies: Communicative Capitalism and Left Politics*. Durham, NC: Duke University Press.

Dean, J. (2010) *Blog Theory: Feedback and Capture in the Circuits of Drive*. Cambridge: Polity Press.

Dean, J. (2012) *The Communist Horizon*. London: Verso.

Dean, J. (2016) *Crowds and Party*. London: Verso.

Deardon, L. (2017) 'Migrant Deaths on Mediterranean Pass 2,000 Mark on World Refugee Day After New Boat Disasters', *Independent*, 21 June, www.independent.co.uk/news/world/europe/migrants-mediterranean-sea-crossings-2000-dead-world-refugee-day-asylum-seekers-boat-disasters-libya-a7800376.html.

Dehghan, S. K. (2017) '8,500 People Lost in Mediterranean Since Death of Three-year-old Alan Kurdi', *Guardian*, 1 September, www.theguardian.com/world/2017/sep/01/alan-kurdi-khaled-hosseini-mediterranean-refugees-sea-prayer.

Deleuze, G. (1992) 'What is a Dispositif', in T. J. Armstrong, *Michel Foucault: Philosopher*. New York: Harvester Wheatsheaf, pp. 159–68.

Deleuze, G. (1995) *Negotiations*. New York: Columbia University Press.

Deleuze, G. and Guattari, F. (1983) *Anti-Oedipus*. Minneapolis: University of Minnesota Press.

Deleuze, G. and Guattari, F. (1987) *A Thousand Plateaus*. Minneapolis: University of Minnesota Press.

Denkee, A. (2000) *Historical Dictionary of Heidegger's Philosophy*. Lanham, MD: Scarecrow Press.

Douzinas, C. and Žižek, S. (2010) 'Introduction', in C. Douzinas and S. Žižek, *The Idea of Communism*. London: Verso.

Dyer-Witheford, N. (2004) 'Species-Being Resurgent', *Constellations* 4(11): 476–92.

Dyer-Witheford, N. (2013) 'Red Plenty Platforms', *Culture Machine* 14.

Dyer-Witheford, N. (2015) *Cyber-Proletariat*. London: Pluto Press.

Eagleman, D. (2015) *The Brain: The Story of You*. London: Penguin.

Edelman, G. M. and Tononi, G. (2000) *A Universe of Consciousness: How Matter Becomes Imagination*. New York: Basic Books.

Elmer, G. (2008) *Preempting Dissent*. Winnipeg: Arbeiter Ring.

Elpidorou, A. (2012) 'Where is My Mind? Mark Rowlands on the Vehicles of Cognition', *Avant* 3(1).

Engels, F. (1978) *Anti-During. In Borodulina, On Communist Society*. Moscow: Progress Publishers.

Fiennes, S. (dir.) (2012) *The Perverts Guide to Ideology* [Film].

Fisher, M. (2009) *Capitalist Realism*. Winchester: Zero Books.

Forbes, E. (2017) 'How Technology Powered the Catalan Referendum', *Open Democracy*, 22 October, www.opendemocracy.net/can-europe-make-it/emeka-forbes/how-technology-powered-catalan-referendum.

Foucault, M. (1980) *Power/Knowledge: Selected Interviews and Other Writings 1972–1977*. New York: Pantheon Books.

Franklin, S. (2009) 'On Game Art, Circuit Bending and Speedrunning as Counter-Practice: "Hard" and "Soft" Nonexistence', *CTheory*, June.

Fuller, M. and Goffey, A. (2012) *Evil Media*. Cambridge, MA: MIT Press.

Galloway, A. (2004) *Protocol*. Cambridge, MA: MIT Press.

Galloway, A. (2012) *The Interface Effect*. Cambridge: Polity Press.

Galloway, A. and Thacker, E. (2007) *The Exploit*. Minneapolis: University of Minnesota Press.

Gazzaley, A. and Rosen, L. D. (2016) *The Distracted Mind: Ancient Brains in a High-Tech World*. Cambridge, MA: MIT Press.

Gill, C. (2010) 'Hijacking of a Very Middle-Class Protest', *Daily Mail*, 11 November, www.dailymail.co.uk/news/article-1328385/TUITION-FEES-PROTEST-Anarchists-cause-chaos-50k-students-streets-html.

Grusin, R. (2010) *Premediation*. London: Palgrave.

Guattari, F. (2000) *The Three Ecologies*. London: Athlone Press.

Han, B.-C. (2017) *Psychopolitics: Neoliberalism and New Technologies of Power*. London: Verso.

Hands, J. (2006) 'Civil Society, Cosmopolitics and the Net: The Legacy of 15 February 2003', *Information, Communication and Society* 9(2).

Hands, J. (2011) *@ is for Activism*. London: Pluto Press.

Hardt, M. (2010) 'The Common in Communism', in C. Douzinas and S. Žižek, *The Idea of Communism*. London: Verso, pp. 131–44.

Hardt, M. (2012) 'Communism is the Ruthless Critique of All that Exists', UniNomade, 3 December, www.uninomade.org/communism-is-the-ruthless-critique-of-all-that-exists/print.

Hardt, M. and Negri, A. (2000) *Empire*. Cambridge, MA: Harvard University Press.

Hardt, M. and Negri, A. (2004) *Multitude*. New York: Penguin.

Hegel, G. (1977) *Phenomenology of Spirit*. Oxford: Oxford University Press.

Heidegger, M. (1962) *Being and Time*. Oxford: Blackwell.

Heidgegger, M. (1966) *Discourse on Thinking*. New York: Harper and Row.

Heidegger, M. (1967) *What is a Thing?* South Bend: Gateway Editions.

Heidegger, M. (1977) *The Question Concerning Technology and Other Essays*. New York and London: Garland Publishing.

Heidegger, M. (2001) *Poetry, Language, Thought*. New York: HarperCollins.

Heidegger, M. (2010) *Being and Time*. Albany: SUNY Press.

Heidegger, M. (2012) *Bremen and Freiburg Lectures: Insight Into That Which Is and Basic Principles of Thinking*. Bloomington: Indiana University Press.

Hofstadter, D. (2000) *Godel, Escher, Bach: An Eternal Golden Braid*. London: Penguin.

Hofstadter, D. (2007) *I Am a Strange Loop*. New York: Basic Books.

Holloway, J. (2002) *Change the World Without Taking Power*. London: Pluto Press.

Honneth, A. (2008) *Reification: A New Look at an Old Idea*. Oxford: Oxford University Press.

Hookway, B. (2014) *Interface*. Cambridge, MA: MIT Press.

Ihde, D. (1983) *Existential Technics*. New York: SUNY Press.

Jobs, S. (n.d.) Keynote Address, www.european-rhetoric.com/analyses/ikeynote-analysis-iphone/transcript-2007.

Jordan, T. (2008) *Hacking*. Cambridge: Polity Press.

Jordan, T. (2015) *Information Politics*. London: Pluto Press.

Jordan. T. (2016) 'A Genealogy of Hacking', *Convergence* 23(5): 528–44.

Kavada, A. (2015) 'Creating the Collective: Social Media, the Occupy Movement and its Constitution as a Collective Actor', *Information, Communication and Society* 18(8): 872–86.

Kellner, D. (1989) *Critical Theory, Marxism and Modernity*. Cambridge: Polity Press.

Khomami, N. (2017) '#MeToo: How a Hashtag Became a Rallying Cry Against Sexual Harassment', *Guardian*, 20 October, www.theguardian.com/world/2017/oct/20/women-worldwide-use-hashtag-metoo-against-sexual-harassment.

Kingsley, P. (2016) 'The Death of Alan Kurdi: One Year On, Compassion Towards Refugees Fades', *Guardian*, 2 September, www.theguardian.com/world/2016/sep/01/alan-kurdi-death-one-year-on-compassion-towards-refugees-fades.

Klein, N. (2007) *The Shock Doctrine: The Rise of Disaster Capitalism*. London: Penguin.

Kojève, A. (1969) *Introduction to the Reading of Hegel*. Ithaca: Cornell University Press.

Kurzweil, R. (2005) *The Singularity is Near*. London: Viking.

Kurzweil, R. (2006) 'Reinventing Humanity', *The Futurist*, March–April: 39–46.

Lazzarato, M. (2006) 'Life and the Living in the Societies of Control', in M. Fuglsang and B. Meier Sorensen, *Deleuze and the Social*. Edinburgh: Edinburgh University Press.

Lazzarato, M. (2014) *Signs and Machines: Capitalism and the Production of Subjectivity*. Los Angeles: Semiotext(e).

Lee, S., So, C. Y. and Leung, L. (2015) 'Social Media and the Umbrella Movement: Insurgent Public Sphere in Formation', *Chinese Journal of Communication* 8(4): 356–75.

Lefebvre, H. (1968) *Dialectical Materialism*. London: Cape Editions.

Lévy, P. (1997) *Collective Intelligence*. Cambridge: Perseus Books.

Ludovico, A. and Cirio, P. (n.d.) 'How We Did It', www.face-to-facebook.net/how.php.

Lukács, G. (1971) *History and Class Consciousness*. Cambridge, MA: MIT Press.

Malabou, C. (2008) *What Should We Do with Our Brain?* New York: Fordham University Press.

Marx, K. (1973) *Grundrisse, Foundation of the Critique of Political Economy*. London: Penguin.

Marx, K. (1976) *Capital, Vol. 1*. London: Penguin.

Mazzucato, M. (2014) *The Entrepreneurial State, Debunking Public vs. Private Sector Myths*. London: Anthem Press.

Medina, E. (2014) *Cybernetic Revolutionaries*. Cambridge, MA: MIT Press.

Meshkit (n.d.) 'Meshkit', www.opengarden.com/meshkit.html.

Metzinger, T. (2009) *The Ego Tunnel*. New York: Basic Books.

Miller, L. (2015) 'The Boston Marathon and Reddit', *Salon*, 15 April, www.salon.com/2015/04/15/the_boston_marathon_and_reddit_when_the_internets_deluded_amateur_hour_detectives_ran_amok.

Mitchell, A. (2012) 'Translator's Forward', in M. Heidegger, *Bremen and Freiberg Lectures: Insight Into That Which Is and Basic Principles of Thinking*. Bloomington: Indiana University Press.

Mitchell, A. (2013) 'The Fourfold', in F. Raffoul and E. S. Nelson, *The Bloomsbury Companion to Heidegger*. London: Bloomsbury, pp. 297–303.

Nagel, T. (1974) 'What Is It Like to Be a Bat?', *Philosophical Review* 83(4): 435–50.

Nagle, A. (2017) *Kill All Normies*. Winchester: Zero Books.

Noë, A. (2009) *Out of Our Heads*. New York: Hill and Wang.

Pasquinelli, M. (2009) 'Google's PageRank Algorithm: A Diagram of Cognitive Capitalism and the Rentier of the Common Intellect', in K. Becker and F. Stalder, *Deep Search*. London: Transaction.

Pinker, S. (1994) *The Language Instinct*. London: Penguin.

Potts, L. and Harrison, A. (2013) 'Interfaces as Rhetorical Constructions: reddit and 4chan During the Boston Marathon Bombings', *SIGDOC' 13*, September 30–October 1.

Rheingold, H. (2002) *Smart Mobs*. New York: Basic Books.

Rowlands, M. (2010) *The New Science of the Mind*. Cambridge, MA: MIT Press.

Shirky, C. (2010) *Cognitive Surplus*. London: Allen Lane.

Silverstone, R. (2003) 'Preface to the Routledge Classics Edition', in R. Williams, *Television and Cultural Form*. London: Routledge.

Simondon, G. (2017) *On the Mode of Technical Objects*. Minneapolis: Univocal.

Singer, P. (1983) *Hegel*. Oxford: Oxford University Press.

Smith, H. (2015) 'Shocking Images of Drowned Syrian Boy Show Tragic Plight of Refugees', *Guardian*, 2 September, www.theguardian.com/world/2015/sep/02/shocking-image-of-drowned-syrian-boy-shows-tragic-plight-of-refugees.

Srnicek, N. (2017) *Platform Capitalism*. Cambridge: Polity Press.

Srnicek, N. and Williams, A. (2015) *Inventing the Future: Postcapitalism and a World Without Work*. London: Verso.

Stiegler, B. (2010) *For a New Critique of Political Economy*. Cambridge: Polity Press.

Stiegler, B. (2013) *What Makes Life Worth Living: On Pharmacology*. Cambridge: Polity Press.

Suarez-Villa, L. (2009) *Technocapitalism: A Critical Perspective on Technological Innovation and Corporatism*. Philadelphia: Temple University Press.

Surowiecki, J. (2004) *The Wisdom of Crowds*. London: Abacus.

Terranova, T. (2000) 'Free Labour: Producing Culture for the Digital Economy', *Social Text* 18(2): 33–58.

The Times of India (2016) '"Firechat" Comes to UoH students' Rescue', 23 January, https://timesofindia.indiatimes.com/city/hyderabad/Firechat-comes-to-UoH-students-rescue/articleshow/50691376.cms.

Tinnell, J. (2015) 'Grammatization: Bernard Stiegler's Theory of Writing and Technology', *Computers and Composition* 37: 132–46.

Torey, Z. (2014) *The Conscious Mind*. Cambridge, MA: MIT Press.

Tufecki, Z. (2017) *Twitter and Tear Gas: The Power and Fragility of Networked Protest*. New Haven and London: Yale University Press.

Turkle, S. (2005) *The Second Self: Computers and the Human Spirit*. Cambridge, MA: MIT Press.

Virno, P. (2004) *The Grammar of the Multitude*. New York: Semiotext(e).

Wark, M. (2004) *A Hacker Manifesto*. Cambridge, MA: Harvard University Press.

Wells, H. (1937) 'Today's Distress and Horrors Basically Intellectual: Wells', *Science News Letter* 32 (861): 229.

Wendling, M. (2018) *Alt Right: From 4chan to the White House*. London: Pluto Press.

Williams, R. (1973), 'Base and Superstructure in Marxist Cultural Theory', *New Left Review*, 82.

Williams, R. (2003) *Television and Cultural Form*. London: Routledge.

Williams, R. (2011) *The Long Revolution*. Cardigan: Parthian Books.

Wrathall, M. (2005) *How to Read Heidegger*. London: Granta Books.

X-Net (n.d.) 'Technopolitics and Hacktivism', https://xnet-x.net/en/areas/technopolitics-hacktivism-artivism.

Zeman, A. (2002) *Consciousness: A Users's Guide*. New Haven: Yale University Press.

Žižek, S. (2002a) 'A Plea for Leninist Intolerance', *Critical Inquiry* 28 (2): 542–66.

Žižek, S. (2002b) 'Introduction: Between the Two Revolutions', in S. Žižek (ed.), *Revolution at the Gates*. London: Verso.

Žižek, S. (2002c) *Revolution at the Gates*. London: Verso.

Žižek, S. (2009) *First as Tragedy, Then as Farce*. London: Verso.

Žižek, S. (2012) *Organs Without Bodies*. London: Routledge.

Žižek, S. (2014) *The Event*. London: Penguin.

Index

The Pluto Press Newsletter

Hello friend of Pluto!

Want to stay on top of the best radical books
we publish?

Then sign up to be the first to hear about our
new books, as well as special events,
podcasts and videos.

You'll also get 50% off your first order with us
when you sign up.

Come and join us!

Go to bit.ly/PlutoNewsletter